IISS

THE IISS

ARMED CONFLICT SURVEY

2016

published by

Routledge
Taylor & Francis Group

LONDON AND NEW YORK

for

The International Institute for Strategic Studies

The International Institute for Strategic Studies
Arundel House | 13–15 Arundel Street | Temple Place | London | WC2R 3DX | UK

THE IISS ARMED CONFLICT SURVEY 2016

First published May 2016 by **Routledge**
2 Park Square, Milton Park, Abingdon, Oxon, OX14 4RN

for **The International Institute for Strategic Studies**
Arundel House, 13–15 Arundel Street, Temple Place, London, WC2R 3DX, UK

Simultaneously published in the USA and Canada by **Routledge**
52 Vanderbilt Avenue, New York, NY 10017

First issued in hardback 2018

Routledge is an imprint of the Taylor & Francis Group, an informa business

DIRECTOR-GENERAL AND CHIEF EXECUTIVE Dr John Chipman
EDITOR Dr Anastasia Voronkova

CONTRIBUTING EDITOR Dr Nicholas Redman
ASSISTANT EDITOR Chris Raggett
EDITORIAL Alice Aveson
DESIGN AND PRODUCTION John Buck, Kelly Verity
CONFLICT AND MAP RESEARCH Raimah Amevor, Daniel Bodirsky, Henry Boyd, Nicholas Crawford, John Drennan, Harry Johnstone, Andrew Kelly, Philemon Leenhardt, Jan Raudszus, Antonio Sampaio, Christopher Shay, Hebatalla Taha, Jeremy Taylor, Hanna Ucko Neill, Jens Wardenaer, Alex Waterman
COVER IMAGES Saleh Mahmoud Laila/Samir Bol/Hamit Huseyin/Anadolu/Getty; Pierre Crom/Milos Bicanski/Getty

British Library Cataloguing in Publication Data
A catalogue record for this book is available from the British Library

Library of Congress Cataloguing in Publication Data

ISSN 2374-0973
ISBN 13: 978-1-138-47429-1 (hbk)
ISBN 13: 978-1-85743-861-1 (pbk)

Contents

Editor's Introduction

The inaugural edition of *The Armed Conflict Survey* charted a steady increase in the lethality of armed conflicts from 2010 to 2014, even as the overall number of active conflicts declined. Results for 2015 indicate that lethality is no longer rising, while the number of conflicts barely changed. Headline numbers, however, can be misleading. The fall in the overall level of conflict deaths for 2015 can be explained solely by a lower figure for Syria, which remained by some distance the most lethal conflict globally, and Iraq. This masked notable rises in fatalities in Afghanistan, Nigeria, Somalia and Yemen, as well as in the drug- and crime-driven conflicts in Mexico and Central America. Moreover, the decade-long rise in conflict-related displacement of people showed no sign of abating.

The Armed Conflict Survey provides an analysis of recent developments in active armed conflicts around the globe, and supplements this with essays on trends and aspects of armed conflict globally. It is perhaps not surprising that the Islamic State, also known as ISIS or ISIL, a principal combatant in Iraq and Syria and a rising presence in other parts of the world, is referenced in all of these essays. Lawrence Freedman identifies the terrorist threat that the Islamic State poses for European states as the main factor prompting Western powers to consider a more muscular engagement in the Middle East, after a period of wariness and self-doubt occasioned by the draining conflicts in Iraq and Afghanistan. Whether they will adopt security policies matching their power capabilities, he argues, will depend partly on the next US president and also on whether policymakers can identify a path through the region's cross-cutting antagonisms and allegiances. Even if they do, the resources committed may fall short of what is needed.

Achim Wennmann addresses conflict from a political economy perspective and notes that, in the case of the Islamic State, it has paid considerable attention to state-building and revenue-raising in order to sustain and protect itself. While Western and Arab airpower has focused on cutting the group's revenue from oil production, refining and trading, Wennmann notes that it has a diversified revenue base including taxation, bank robbery, kidnapping and foreign donations. That diversity makes Islamic State a more durable entity from a funding perspective than some observers might assume.

Daniel Byman focuses on foreign fighters and the potential for blowback in Europe. He notes that there are more foreign fighters active in Syria and Iraq than in the anti-Soviet conflict in Afghanistan, which in turn featured more foreign fighters than the subsequent conflicts in the Balkans, Chechnya and post-9/11 Afghanistan and Pakistan combined. Yet he cautions against assuming that this will translate into a wave of terrorist attacks in Europe for numerous reasons, including improvements in the capability and vigilance of western security services over the past 15 years.

While all of the authors touch on the Islamic State, their essays also address broader concerns. Carol Mann charts the increase in female armed militancy, which is partly a consequence of technological changes and the fact that the modern battlefield can be a high street or hospital. She looks at the roles assigned to women in different kinds of insurgent movements and state armed forces, their motivations and the unique challenges facing them in the conflict and post-conflict phases. Social injustice and marginalisation drive the recruitment of women in many contemporary conflicts, she argues, adding that despair may supplant ideology as the principal motive for action.

Wennmann guides us through the evolution of our understanding of the drivers of war, from a focus on identity to economic motives and more recently to multi-factor explanations that include socio-economic factors, politics, ethnicity and security dilemmas, some of which may be narrowly local in nature. Understanding the drivers of violence and structure of armed groupings is important, he argues, for effective policy. In the case of Mexico, state efforts to decapitate and fracture cartels have increased insecurity rather than decreased it. Freedman, recalling the 1990s Revolution in Military Affairs (RMA), observes that precision-strike capabilities have increased Western options against irregular forces, but when these blend with civilians in population centres the efficacy of such advanced capabilities is largely dependent on having accurate, timely intelligence.

Dynamic uncertainty

Throughout 2015 insurgents, militia fighters and other armed actors continued to use a mixture of conventional and asymmetric tactics, often employing IEDs to kill or injure large numbers of people, particularly civilians. From a military perspective, some of the most revealing conflicts were those in Somalia, Afghanistan, Ukraine, Syria and Iraq. Operations in Somalia demonstrated the growing capabilities of African forces operating under the command of the African Union. Despite experiencing various setbacks, they made slow but steady progress in stabilising the country. This reflected the determination of African states to improve their military capabilities, and to deal with security challenges on the continent themselves. While the United States and the European Union provided a range of political, developmental and military training and advice – and conducted special-forces raids and strikes using unmanned aerial vehicles – African soldiers fought the major battles on the ground.

Following the drawdown of US and NATO troops in Afghanistan, the losses sustained by Afghan forces suggested weaknesses in command, control, communications and logistics. They also showed the difficulty of building combat units during a war against a resilient enemy, especially in a country with underdeveloped human capital and significant ethnic and political fault lines, as well as deep-seated patronage, warlordism and corruption. These factors not only limited the effectiveness of the Afghan army and police, but also made the development of the Afghan Air Force particularly problematic.

Despite being predominantly conventional, with largely static trench lines, the conflict in eastern Ukraine often involved ceasefire violations that resulted in high casualties. The majority of deaths and injuries among government forces resulted from the separatists' artillery strikes, which benefited from signals intelligence and surveillance using unmanned aerial vehicles. Together with the decisive role played by Russian artillery in stopping the Ukrainian government's autumn 2014 offensive, this suggested that artillery was again becoming not only a decisive factor on the battlefield, but also a major cause of military and civilian casualties.

Artillery, along with mortars and rocket launchers, also played an important role in the linked conflicts in Syria and Iraq. The war in Syria defied straightforward analysis, becoming even more complex as the Syrian government, Hizbullah and other Iranian-backed forces battled Jabhat al-Nusra, Kurdish fighters, a variety of other opposition groups and the Islamic State. In both Syria and Iraq, ISIS employed tactics that included elements of a conventional combined-arms approach, albeit with innovations such as the use of makeshift combat engineer vehicles to breach the defences of government forces. Meanwhile, the slow rebuilding of the Iraqi army by the US and its allies involved the development of conventional capabilities and tactics. These were essential in the effort to recapture Ramadi, as seen in the deployment of Iraqi tanks and combat engineer vehicles there.

Coordinating with Iraqi troops and Kurdish forces, the US-led coalition made use of airpower, deployed with strict rules of engagement and considerable precision, to help constrain the growth of ISIS. Russian airpower, applied with far less precision and accompanied by a dissimulating information campaign, helped the forces supporting the Syrian government halt the deterioration of their position and launch an operational-level counter-offensive. Despite taking different forms, both air campaigns indicated the importance of integrating and coordinating air and land forces.

The kaleidoscope of politics, fighting and information operations in Syria and Iraq seemed to contain many examples of hybrid warfare, mixing state and non-state actors, as well as symmetric and asymmetric capabilities. The elaborate developments in the countries reflected the unpredictability of war – where the enemy has a vote, and may fight to the death to cast it. They also confirmed that war is a clash of wills between actors seeking to shape events to suit their political aims, exploiting any advantage they can. In the dynamic uncertainty of the hybrid security environment, one can assess the dominant trends, threats and challenges in the conduct and evolution of such conflicts, but many questions about their future development remain unanswered.

Chapter One: **Thematic Essays**

Pessimism and Planning in Armed Conflict

Lawrence Freedman

Predictions about developments in inter-state and intra-state armed conflicts are naturally influenced by recent experiences. This explains the optimism that followed the end of the Cold War and the singular experience of the 1991 Gulf War, and the pessimism now following the long insurgencies in Iraq and Afghanistan, the souring of relations with Russia, the anxieties surrounding China, and the sudden prominence of the Islamic State (also known as ISIS or ISIL) as a source of mayhem in the Middle East and terrorism in Western countries. The current pessimism demonstrates why the former optimism was overdone. Might the reverse also be true, in that the current gloom understates the ability of Western countries to manage international affairs more effectively and reduce threats?

From overconfidence to self-doubt

The optimism of the 1990s reflected relief at the end of the Cold War and the successful handling of the Iraqi invasion of Kuwait. Other matters were handled less well, as seen in the civil war in Rwanda and the break-up of Yugoslavia, but there were successes in humanitarian intervention and concepts of best practice were under development. Much of the optimism reflected American predominance, with talk of a unipolar world. This could be seen as a benign hegemony, under which alternative ideological dispositions were cowed by the failure of European communism and the success of the liberal capitalist model that others wished to emulate. This predominance was to be partly sustained by the so-called Revolution in Military Affairs (RMA), which was being driven forward by a mixture of new technologies that together made it possible to apply force with speed and precision, and to maximum effect – thus allowing for decisive victories against 'peer competitors'. Much of the weaponry associated with that vision is now in place. Fortunately, we have not needed to find out whether the equipment will work well in the contingency for which it was designed: a large-scale war between major powers. Instead, it has been in use in contingencies that were taken far less seriously in the 1990s: counter-insurgency and counter-terrorism.

These activities have borne little resemblance to the sort of high-tempo and almost bloodlessly decisive regular warfare envisaged by the RMA. If wars could be fought according to the dictates of the RMA, few opposing armies would stand much of a chance. Opponents organised conventionally for a classical battle, such as the Taliban in 2001 or Iraqi forces in 2003, were overwhelmed, but it then proved to be far more difficult to support replacement governments in the face of constant harassment and attack from insurgents deploying improvised explosive devices, mortars and snipers, as well as light trucks carrying machine guns for mobile operations. In these messier fights, with the enemy moving in and out of populated areas, the lines between civilians, militias and regular forces become blurred. Western air power can strike with extraordinary precision but only so long as the targets have been accurately identified in the first place, especially given the need to limit the impact on civilian life and property. Drones engaged in targeted killings demonstrate how new technologies can create new tactical possibilities, even within counter-terrorism, but at the same time they also raise significant ethical and legal questions, and there are limits to the extent to which a determined movement with a tendency for martyrdom can be brought down by picking off members of its leadership cadre.

Despite apparently overwhelming military superiority, Western armed forces have struggled in many recent conflicts, leading to a loss of confidence in their ability to apply military force successfully on the ground. This loss has been evident in the reliance on air power in Libya, Iraq and Syria, leaving any ground action to locally generated forces, which are part of the political struggle at the heart of the conflict, and whose objectives may be only temporarily and occasionally aligned with those of the Western countries engaged.

One answer to this pessimism is to point out that there may still be a deterrent effect at work. The West's enemies and potential peer competitors are unwilling to consider conventional warfare precisely because they know it is a losing proposition, just as all the major powers understand that they should not risk nuclear war because of the prospect of mutual assured destruction. Russian President Vladimir Putin, for example, has been ready to invade bordering countries that are not in NATO but he has been more circumspect when it comes to Alliance members with which Russia shares a border. He has sought to deter NATO from coming to the aid of Ukraine not so much by emphasising Russian conventional strength but by drawing regular attention to the country's nuclear capabilities. There are worries about a major war that flow from Russia's assertiveness in Europe and China's in Asia, but there is no suggestion that either of these countries is sanguine about the prospect. Both pick their targets with care, pushing and probing while trying to avoid over-commitment, perhaps waiting for a moment when the situation might favour them more.

For the moment, the challenge with Russia and China is, in different ways, to address their occasional assertions of power with measures that avoid the risk of direct confrontation but deny them the advantages of being able to act with impunity. With regard to Ukraine, for example, this has involved economic sanctions and helping Kiev develop its own means to deal with the separatist strongholds in the eastern part of the country. With China's efforts to assert its position in the South China Sea, this has meant physical demonstrations that its claims are not accepted. It is unclear whether these measures will suffice in the long term, especially given the assumption that Chinese strength is destined to grow over this century.

Destructive complexity in the Middle East

There are other current challenges, such as the tension between India and China, and the real dangers posed by a nuclear-armed and erratic regime in North

Korea. But the most pressing immediate problem is how to deal with the chronic instability in parts of the Middle East and North Africa, and the growing role of Islamist extremists. On the basis of past experience, a compelling response to this challenge has appeared to be stay clear, recognise the risks of intervention and accept that little can be done to support friendly governments, change dubious regimes or rescue vulnerable sections of their populations from persecution. This was the initial response to the developing civil war in Syria, as President Bashar al-Assad responded to demands for reform with harsh action against rebellious towns and cities. Even when chemical weapons were used in 2013, US President Barack Obama and UK Prime Minister David Cameron found that their electorates remained wary of any engagement. Estimates vary, but up to 300,000 people may now have been killed as a result of this war.

What really made the difference in terms of Western attention was the rise of ISIS. Under its original guise as an al-Qaeda franchise, it pursued a murderous campaign after the 2003 invasion of Iraq until 2006, when its leader, Abu Musab al-Zarqawi, was killed and the group alienated its potential supporters in the Sunni community. It was forced out of its strongholds, but it eventually managed to return via Syria. Taking advantage of the civil war, it took over a chunk of Syrian territory and then, in 2014, swept back into Iraq, pushing aside a weak Iraqi army that appeared reluctant to defend Sunni areas. Only as ISIS approached Baghdad and Kurdish territory did it meet with serious resistance, bolstered by the return of Western air power to Iraq.

Ruthlessness and control of land straddling Iraq and Syria allowed ISIS to make some progress towards its goal of establishing a functioning state, or 'caliphate', from where authority might be claimed over the whole of the Muslim world. Groups swearing allegiance to ISIS can be found in Libya, still seeking stability after the overthrow of the Gadhafi regime in 2011, and in other unsettled states of Africa. Although Assad's forces remain largely responsible for the carnage in Syria, ISIS has been a contributory factor (not least by complicating the anti-Assad struggle). One symptom of the region's instability was a surge of desperate people making their way from Syria and Iraq to Europe, putting enormous strain on the open internal borders of the European Union. Another was a growing incidence of terror attacks in Europe and beyond. Initially, the acts were largely perpetrated by self-starting ISIS sympathisers. Many, though not all, were thwarted. Then the attacks began to demonstrate planning and strategic direction, culminating in November 2015 with two coordinated, substantial and deadly operations – the bombing of a Russian airliner flying out of Egypt was followed by multiple assaults in the centre of Paris.

In the first instance, this was addressed as a problem of domestic counter-terrorism, involving issues such as visa checks, airport safety and gun control, as well as challenging questions of how best to integrate minority communities and implement proper levels of state surveillance. It is unrealistic to expect to prevent all attacks, especially those by 'lone wolves', but their political effect is manageable if they remain rare and are only an intermittent source of disruption. The key requirement is to prevent the development of a persistent and irresistible campaign. In particular, to frustrate the more substantial attacks, terrorists must be denied bases at which militants can be trained and equipped and their operations organised. The thought of a tract of vital territory so close to Europe being controlled by such an extreme group provided the rationale for a Western campaign against ISIS.

But if ground troops (other than special forces) were excluded, there were limits on what could be achieved by air power alone. The British debate on air-strikes against ISIS in Syria, for example, involved demands to identify the rebel forces that might be available to provide the ground component of the campaign. The 'moderate' forces identified were a mixed bag, many not natural allies of the West and most at that time preoccupied with their fight against Assad. It soon became apparent that without this component there were only a few targets that could usefully be attacked form the air. It might be possible to prevent ISIS from expanding further using a combination of local forces and Western air power, but the group's territorial gains will be much harder to roll back.

President Obama described US objectives as being 'to degrade and ultimately destroy' ISIS.[1] The problem with 'ultimately destroy' is that ISIS is only the latest in a line of extremist groups with a shared ideological root and support that knows no national boundaries. There are significant differences among Islamist groups, which matter to their leaders and occasionally to their followers, but they draw on similar themes. They are often in competition for recruits from the same broad constituency that embraces radical policies based on an exclusive and austere view of Islam, and has a conviction that the faith must be defended and promoted through violent action. It does not depend on a particular state or charismatic leader, which is why comparisons between Islamism and the 'isms' of fascism and communism are unhelpful.

President Obama's objective of degradation could imply an attritional strategy, seeking to kill off as many of the enemy as possible. If so, it would risk falling victim to the 'body count' fallacy that became notorious during the Vietnam War, when the innocent dead were counted as fighters, and the capacity of the enemy to replenish its numbers was underestimated.[2] If degradation can be of real strategic value in this context, it has to be about undermining the concept of the caliphate, by reducing the amount of territory ISIS controls, and so denying it a base.

Such an objective means addressing the organisation's ability to support the populations under its control, by disrupting its financial and trafficking networks, as well as by harassing its leadership and reducing its capabilities. There are always internal tensions in groups like this, which is why they often fracture and begin to fight amongst themselves, and that could be encouraged. The most helpful development, however, would be for local forces to take on ISIS sufficiently to expel it from the territory it currently holds. For this to happen, there has to be more conflict resolution, reform and reconciliation in the region, especially in Syria.

Violence cannot be calmed nor reconstruction begin without major political changes, yet there is little agreement on what these changes should be, and those who would expect to lose as a result will fight hardest to prevent them. The idea that, after a point, people become so exhausted that they readily embrace deals that would have been previously rejected is belied by the experience of countries such as Sudan or the Democratic Republic of the Congo, where conflict has become a way of life and many are invested in the local power and criminal opportunities associated with continuing violence.

There might be an anti-ISIS consensus among members of the UN Security Council, if on little else, but attempts to develop a common policy on Syria soon come up against the complex political currents of the Middle East. The region is beset by an extraordinarily intricate set of cross-cutting antagonisms and allegiances, to the extent that it becomes almost impossible for any actor, whether from within or without the Middle East, to deal with one set of problems without at the same time aggravating others. Every move to weaken one opponent has risked strengthening another. This has been as true for Russia as for the United States – as true for Iran as for Saudi Arabia.

This mixture of interests and priorities makes it impossible to concoct a united front against ISIS. Some in the West took the view that making the fight against ISIS a priority meant taking the pressure off Assad, though he could never again preside over a united country. The local rebel militias capable of acting against ISIS had as their first priority toppling Assad. When President Putin escalated Russia's engagement with Syria in September 2015, his main priority was to sustain Assad, and this led to a definition of terrorists that extended to all the Syrian president's opponents and downgraded the relative importance of ISIS. Iran was closely involved in the fight against ISIS in Iraq but in Syria, along with its Lebanese allies in Hizbullah, it was invested heavily in Assad. Members of the Gulf Cooperation Council, particularly Saudi Arabia, recognised the danger in ISIS but were more fearful of what they saw as Iran's malign regional influence. They saw the war in Syria as part of a wider Sunni–Shia struggle. At the start of 2016, Saudi Arabia executed a leading Shia cleric, prompting the sacking of its Embassy in Tehran, and leading to a break in diplomatic and most economic relations between Iran and the Gulf Arab states. The latter might have a common interest with Israel in opposing Iran and its allies, but they could not take this potential alliance too far because of the Palestinian issue. Israel had more or less stable relations with Syria under Assad and his father after the 1973 war, and had no direct interest in his overthrow. Meanwhile, Turkish President Recep Tayyip Erdogan, hostile to ISIS but even more so to Assad, was fearful of encouraging Kurdish autonomy. At the end of November 2015, he had a spectacular falling out with President Putin after Turkish forces shot down a Russian aircraft. He blamed the incident on the aircraft's violation of Turkish airspace, but also criticised Russian attacks on a Turkoman group fighting Assad's army.

Strategies of caution

As different parties sought to navigate their way through this morass, they were as likely to stumble and face setbacks as they were to find a clear path. The challenges precluded quick fixes, resulting in stalemates or else progress through attrition and incremental steps, with strategies of hedging, marked by resources held in reserve, conditional alliances and qualified commitments. Even the most powerful of states were unsure they could get a grip on the situation.

Yet few felt they could ignore it. For Western countries, the events of the past few years have demonstrated that they cannot avoid exposure to some security risks – whether these are faced domestically or as a result of attempts to address threats at the source. The lessons are that they cannot insulate themselves from conflicts in their neighbourhood; that they have to start to make better provisions for their security, looking hard at how they collect intelligence and develop military capabilities; and that they need to start thinking again about how to respond to fractured states and the disruptive political movements that they spawn. Diplomacy has become more active and military operations have begun, albeit cautiously.

At the same time, the countries that were supposedly taking advantage of Western passivity are facing troubles of their own. If a vacuum has been created by President Obama's evident distaste for high-risk military action, others are struggling to fill it and have yet to show a sure touch. President Putin discovered that it was one thing to catch your rivals unawares with bold moves but quite another to sustain forward positions in Ukraine and Syria over time, especially with a weakening economy. Russia's economy has been rocked by the collapse in the oil price and hampered by sanctions, causing it to plummet from sixth in the world to 16th. Iran, with a fragile economy also affected by the low oil price, accepted constraints on its nuclear programme in exchange for the reduction of international sanctions. Saudi Arabia's own interventions in Syria and Yemen have yet to produce the desired results. Turkey has lost its swagger as it faces a renewed Kurdish insurgency and tense relations with Russia. Even China, which has largely distanced itself from turmoil in the Middle East, has suffered major setbacks in its economic management. Elsewhere, former emerging economies, such as Brazil, have slipped back while the radical socialist experiment in Venezuela has been a calamitous failure.

Trends mislead and choices matter. Countries with considerable economic, political and military strength can be tempted into crude assertions of power without thought for the long-term consequences, leading to the loss of capacity and confidence. After a series of bad experiences, Western states are displaying understandable caution and they are not likely to take on more risks just because potential rivals with different economic and political models are having their own difficulties. But they are now being forced to engage with sets of problems they might have preferred to ignore and have been drawn back into countries from where they tried to disengage. Having given the impression of punching below their weight for some years, they are now starting to accept the need for security policies that begin to match their underlying economic strength, including looking to repair their defences. Whether a turning point has been reached will depend in part on the outcome of the American presidential election, but also on whether any effective strategies can be identified to help them navigate their way through the multiple challenges faced, especially in the Middle East. Even with a greater determination to shape events on the ground, the gap between what they are prepared to commit and what the situation may demand is substantial.

Lawrence Freedman is Emeritus Professor of War Studies at King's College London and author of *Strategy: A History* (Oxford University Press, 2013).

Notes

1 White House, 'President Obama: "We Will Degrade and Ultimately Destroy ISIL"', 10 September 2014, https://www.whitehouse.gov/blog/2014/09/10/president-obama-we-will-degrade-and-ultimately-destroy-isil.

2 The US unmanned aerial vehicle programme counted 'men of military age' in a target area as combatants. Jo Becker and Scott Shane, 'Secret "Kill List" Proves a Test of Obama's Principles and Will', *New York Times*, 29 May 2012, http://www.nytimes.com/2012/05/29/world/obamas-leadership-in-war-on-al-qaeda.html.

The Political Economy of Violent Conflict

Achim Wennmann

With violent conflict rapidly changing the strategic landscape in many parts of the world, new perspectives are required to inform analysis and policy responses. This essay focuses on how a political-economy perspective could provide policy-makers with a means to better understand the organisational aspects of armed groups and the behaviour and motives of their members, often vaguely labelled 'terrorists', 'warlords' or 'criminals'.[1] The practical value of this perspective rests on delineating the functions of violence, the characteristics of armed groups and the financing of conflict – all of which help to develop a better understanding of conflict dynamics in fluid and turbulent circumstances. These perspectives can build on over two decades of work on the political economy of violent conflict.[2]

The changing strategic landscape

'We have entered a new era, and it is not a peaceful one', remarked Peter Maurer, president of the International Committee of the Red Cross (ICRC), in October 2015.[3] Maurer was discussing the effects of conflict on civilians, in a statement issued with UN Secretary-General Ban Ki-moon – the first joint appeal by the UN and ICRC 'to states and other actors engaged in armed conflict to respect and protect the principle of humanity'.[4] The statement was also a rare public acknowledgement of the ways in which the strategic landscape is changing in many parts of the world, leaving behind the post-Cold War period and perhaps even an era shaped by terrorism and anti-terror policies.

The contours of what is coming may still be blurred, but some elements are becoming clear. While the confrontations in Syria, Ukraine and Yemen act as a sharp reminder that inter- and intra-state wars remain a part of a new strategic landscape, they do not tell the whole story: the majority of conflict-related situations do not fall neatly into either category.[5] A recent report by the Geneva Peacebuilding Platform noted that, in many regions, the security environment is transforming due to 'chronic political instability, persistent social volatility and conflict, the proliferation of non-state armed groups and transnational actors, disputes over land and natural resources, weak state systems, and recurring cycles of violent competition over the state or markets'. The US Department of Defense estimates that climate change will have wide-ranging security implications because 'it will aggravate existing problems ... that threaten domestic stability in a number of countries'.[6] Furthermore, cities are becoming an ever more important space for conflict, crime and violence.[7]

Of course, observations about the changing strategic landscape have been repeatedly made in the forecasts of the last three decades.[8] But there may be something different in the confluence of factors reaching what writer Malcolm Gladwell calls a 'tipping point' – 'that one dramatic moment when everything can change all at once'.[9] This may have already occurred in North Africa and the Middle East; similar tipping points may be reached in fragile states in Latin America, Asia and Africa sometime in the next decade, as societies experience heightened tension. Yet, due to the uncertain timeline for diverse problems converging in a crisis, these developments may take on the characteristics of 'slow onset emergencies' that do not 'emerge from a single, distinct event but [emerge] gradually over time, often based on a confluence of different events', as the Office for the Coordination of Humanitarian Affairs puts it.[10]

As policymakers shape responses to the new strategic landscape, it is useful to look back before looking forward.

Perspectives on conflict dynamics

In the late 1990s, academics and policymakers developed a heightened interest in the political economy of civil wars.[11] During this period, growing access to global markets transformed war economies by allowing for local resources to be sold across the world, and by facilitating the supply of know-how, man-power and war-fighting materiel to theatres of conflict.[12] In these 'new wars', violent conflict was 'not simply a breakdown of a particular system, but a way of creating an alternative system of profit, power and even protection', as scholar David Keen put it.[13] Civil wars were increasingly protracted and self-financing; they were not purely political contests between the parties but 'economics by other means' – as observed by historian Martin Van Creveld, adapting Carl von Clausewitz – in which the objective was often to prolong the fighting for economic gain.[14] In academia, the political economy of violent conflict was recognised as a distinct strand of security studies, and became part of an effort to broaden the discipline beyond state- and military-centred approaches.[15]

As part of this trend, the role of natural resources in civil wars gained particular prominence. Detailed reports by the UN Sanctions Monitoring Mechanism on Angola, Sierra Leone and the Democratic Republic of the Congo were used to name and shame governments and businesses profiting from war economies.[16] A variety of advocacy groups worked to expose the wartime deal-making between governments and rebel groups, as well as their corporate or state supporters.[17] Over time, scholars developed an ever more detailed understanding of the ways in which specific conflicts and broader conflict dynamics were linked with natural resources such as oil, diamonds, timber and poppies.[18] Overall, access to natural resources provided some rebel groups with the means and incentive to challenge weak states.[19]

Academics also attempted to find alternative explanations for armed conflicts once viewed as being driven by ethnic and cultural identities.[20] The debate shifted the focus from viewing conflict as a product of identity, an approach popular during the Balkans wars of the late 1990s, to viewing it as driven by greed, in light of a series of civil wars in Africa, resulting in a debate that became known by the label 'greed and grievance'. Much statistical research emphasised the importance of the feasibility of organising a civil war rather than the motivations for doing so.[21] The debate led to divisions between scholarly disciplines with differing methodologies, especially those relying on either quantitative or qualitative techniques.[22] According to William Zartman, the debate lined up 'true believers' and 'true doubters', until a middle ground was found several years later.[23] That middle ground became studying the role of economic factors in the dynamics of specific conflicts,[24] and in terms of how they interacted with 'socio-economic and political grievances, interethnic disputes, and security dilemmas in triggering the outbreak of warfare', as Karen Ballentine wrote.[25]

Overall, the debate on conflict dynamics has shown that armed conflicts can rarely be explained by a single factor – in how they start, evolve or end. As such approaches remain popular in policymaking circles – a fact illustrated recently by the emphasis on the religious divide between Shi'ites and Sunnis as a reason for conflict in the Middle East – it is important to recall that there is a strand of scholarly work that debunks single-factor explanations of violent conflict. In this context, a political-economy perspective emphasises the importance of local drivers of conflict, which are often determined by messy, contradictory and illiberal realities of specific contexts.[26]

Functions of violence

Viewing conflict from the perspective of political economy also underlines the importance of understanding the functions of violence. Acts of violence have a wide variety of motivations, and may perform psychological, political, social and economic functions.[27] Facing a close election, a local politician might hand out 'scotch and stones' to a loyal youth group, in an effort to keep opposition voters at home. On the individual level, armed violence has often been associated with feelings of impunity, excitement or self-aggrandisement.[28] Such a functional conception rests on an understanding of violence as the result of purposefully organised actions that have strong local dimensions. Even if violence is part of a large national or international conflict, the motivations for violence in a specific context can be highly local, as demonstrated by a strand of research on the micro dynamics of violent conflict.[29] As argued by David Keen, underlying this 'rational kind of madness' are questions such as 'what use is violence? What functions does it assure? In what strategies is it integrated?'[30] In an era in which the media mainly depicts violence in an emotive manner using shocking images and videos, these questions rarely appear in the headlines. Yet they are important for understanding the dynamics of current conflict settings ranging from armed conflict in Syria to election-related violence in Burundi or gang wars in El Salvador.

Al-Qaeda's use of hostages is a case in point. Each execution of a hostage by the group is presented in the media as a barbaric act – which surely it is. Yet the taking and killing of hostages has a crude economic rationale. Writing in the *New York Times* in July 2014, journalist Rukmini Callimachi observed that 'only a minority of hostages – 15 percent ... have been executed or have died since 2008, several of them in botched rescue operations'.[31] Al-Qaeda has largely selected hostages from countries whose governments are known to pay ransoms, and the majority of hostages killed by the group have been from those whose governments are known not to pay. In the words of a man whose brother was taken hostage and killed by the group, 'a UK passport is essentially a death certificate'.[32] Al-Qaeda's revenue from hostage-taking is estimated to have been between US$125 million and US$165m in 2008–14, with payments made 'almost exclusively by European governments'.[33] In this context, hostage killings can pressure governments into paying larger ransoms, or can send an ideological message to a group's followers at relatively low cost. To improve the effectiveness of its hostage-taking, al-Qaeda even implemented subcontracting arrangements for distinct segments of the process.

Organisational structure of armed groups

The functional optics of violence suggest that it is important to grasp the organisational characteristics of armed groups. From the perspective of political economy, armed conflict is the organisation of violence at various orders of magnitude. This means that such conflict is dependent on both its organisers, whose actions exploit and change conditions on the ground, and the material and intangible resources that allow them to do so. A conflict organiser transforms grievances into collective violence, and acquires the resources to initiate and sustain such violence.[34] In this respect, the frequent depiction of the Islamic State (also known as ISIS or ISIL) as a terrorist organisation has misrepresented the nature of the group. Controlling large stretches of territory in Syria and Iraq, including major cities and many regional transportation hubs, ISIS is a well-developed organisation comparable to Colombia's FARC or Sri Lanka's Liberation Tigers of Tamil Eelam (LTTE). According to the ICRC, by late 2015 around 10m people lived in areas controlled by ISIS.[35] In December 2015, leaked ISIS planning documents detailed the extent of the group's strategy for building a civil service, regional governments and economic infrastructure. Commenting on these documents, US general (retd) Stanley McChrystal said that 'if the west sees Isis as an almost stereotypical band of psychopathic killers, we risk dramatically underestimating them'.[36] As argued by Charlie Winter and other analysts, 'far from being an army of irrational, bloodthirsty fanatics, [ISIS] is a deeply calculating political organisation with an extremely complex, well-planned infrastructure behind it.'[37] Thus, in looking at the organisational aspects of armed groups and their military strategies, a functional perspective can help cut through the layers of misleading discourse surrounding such organisations.

By focusing on the organisational aspects of armed groups, analysts and policymakers can also help formulate effective policy responses to 'conflict entrepreneurs', which the United States Institute of Peace defines as 'any group or individual whose profits depend on conditions that promote conflict'.[38] One of the reasons that armed groups build strong organisational networks is to reduce their dependence on any individual. This process is comparable to the creation of permanent state bureaucracies that conduct statecraft and deliver services to the population despite periodic changes in the political leadership. In a highly centralised rebel group, the death or capture of its leader has a profound impact on the organisation, as exemplified by the killing of Jonas Savimbi of the National Union for the Total Independence of Angola (UNITA), or the capture of Shining Path leader Abimael Guzmán in Peru. But such groups are the exception, especially in an era in which war zones are fully integrated with the global economy.[39]

Yet responses to civil wars that focus on individual conflict entrepreneurs remain popular with politicians who wish to demonstrate that they are taking action, and to have a significant effect on the media, especially under the banner of 'countering violent extremism'. The Bureau of Investigative Journalism estimates that by mid-2015 the United States had conducted at least 646 strikes using unmanned aerial vehicles in Pakistan, Somalia and Yemen, killing at least 5,127 'enemy combatants'.[40] Yet, according to journalist Scott Shane, the remote operators carrying out the operations 'often do not know who they are killing, but are making an imperfect best guess'.[41] It is estimated that, as a result, at least 22% of those killed were civilians.[42] And beyond this cost in lives, there remain serious questions about how such targeted killings fit into an overall strategy of de-radicalisation, especially if they only lead to the constant reconfiguration of armed groups, further radicalisation of their existing members and a boost to their recruitment efforts. These effects have been mirrored in the so-called 'kingpin strategy' of Mexico's war on drugs, initiated during the presidency of Felipe Calderón. The strategy centres on a targeting method designed to eliminate the command-and-control elements of drug-trafficking organisations.[43] In Mexico, the killing of cartel leaders has weakened and fractured these groups, but it has also led to spirals of violence that the government could not control, as well as to a new dynamic in which the cartels reconfigured. Mexico's then-attorney general Jesús Murillo Karam said that the kingpin strategy 'brought the lieutenants into command, who were typically more violent – the most capable killers, to say it plainly. They began to take over or form their own groups and the displaced groups began shifting to other types of crimes, and that's where we see kidnapping, extortion, and protection rackets.'[44]

In the difficult task of designing policy responses to violent conflict, the perspective of political economy can explain the resilience of armed groups by emphasising their organisational characteristics. Thus, eliminating one conflict entrepreneur in an organisation whose operations are embedded in strong networks is likely to have little practical effect on the organisation itself. And if those caught in the crossfire happen to be moderate leaders from within or outside the organisation, de-radicalisation initiatives will remain elusive.[45]

Financing of armed groups

Policy responses to violent conflict often target not only the leadership of armed groups, but also their sources of finance. As mentioned above, in the late 1990s and early 2000s, those studying the political economy of violent conflict focused much of their attention on the role of natural resources in conflict financing. Arguably, this approach has been unhelpful for understanding such financing more broadly, as there is a need to capture the broader forces at work in conflict economies.[46] The bias towards natural resources emerged from the characteristics of warring parties in African civil wars at the time – particularly those in Angola, Sierra Leone, Liberia and the Democratic Republic of the Congo. The recommended policy responses included a mixture of sanctions, such as financial measures and travel bans; naming and shaming perpetrators; multi-stakeholder commodity regimes such as the Kimberley Process; and military operations in areas rich with natural resources. Overall, the analysis of conflict financing was relatively straightforward: if an armed actor – either state or non-state – depended on one kind of natural resource or other source of income, it would become vulnerable when it lost access to this source. Yet diversified financing and the ability to adapt to shortfalls in income determine an armed group's resilience. For example, UNITA depended heavily on diamonds and was therefore vulnerable to diamond sanctions and advances by government forces in diamond-mining areas.[47] In Sri Lanka, in contrast, LTTE supplemented its income from diaspora supporters by drawing on civilian economic activity in the territory under its control. The group also invested in legitimate businesses in Asia after 2003, to further reduce its dependence on the diaspora.[48]

There are many other ways to finance armed groups, including through centralised war economies, external assistance from governments, asset transfers from civilians, currency forgery, protection rackets, landing fees and kidnapping. As was widely reported in late 2015, ISIS maintains a complex financing machine, which encompasses oil production and trade, local taxation, bank robbery, kidnapping, extortion and donations from private individuals.[49] As a consequence, it has been described as 'the richest and most destabilising jihadist group of the past 50 years', in the words of journalist Shiv Malik.[50] Estimates of the revenue generated by these sources of income have varied greatly. In November 2014, several news agencies cited sources in German intelligence as saying that the oil revenue generated by ISIS was less than US$100m, significantly lower than the estimate of US$1bn made by US Central Command and figures of between US$2bn and US$3bn put forward by other experts. The estimate made by German intelligence was based on ISIS oil production in Syria of 28,000 barrels, 10,000 of which would be trafficked to neighbouring countries and sold at a price of US$25 per barrel – far below the then-market price of US$80.[51] Unlike previous estimates, the study by German intelligence calculated for the discounted price, the use of oil in warfare, the costs of operating captured facilities and the group's supply of oil to civilians. Moreover, by November 2014, ISIS had lost control of more than 50% of both the oil-production facilities it once held and of the qualified personnel that ran oil operations. Yet despite these losses, the group remained financially strong, having amassed an estimated US$700m–US$2.4bn through bank robberies and taxation.[52] These reports on ISIS once again demonstrated the difficulty of estimating an armed group's financial health amid an ongoing war.

Analyses of conflict financing tend to be one-sided from a military–strategic perspective. It makes a difference whether billions of dollars in revenue belong to 'psychopathic monsters', 'barbarians' and a 'death cult' – as ISIS has been described by US Secretary of State John Kerry, French President François Hollande and British Prime Minister David Cameron respectively – or to a de facto state army that must hold and defend territory and cities through conventional warfare, in the face of air and ground attacks.[53] Overall, most analyses of conflict financing obsess over how much money a group has, and ignore how much it needs – and for what. There are tools for comparing the availability of revenue with the cost of mobilisation for armed groups,[54] based on the premise that what matters in conflict financing is that revenue meets the requirements for implementing a military strategy and achieving a particular objective.

Tackling the dynamics of violent conflict

On 18 December 2015, finance ministers from the 15 countries represented on the UN Security Council adopted a resolution to curb the financing of ISIS.[55] For those who recalled similar efforts against rebel groups in Africa, the meeting underscored the enduring appeal of policy measures that target conflict financing. Yet as David Keen remarked when such efforts first gained traction 20 years ago:

> A focus on the costs of war only takes us so far. Unless we look also at the functions of war, it is difficult to see how wars can be brought to an end … If we try to assess who is benefiting from war and in which ways (as well as who is suffering from war and in which ways), we will be well placed to suggest policy innovations that will alter the balance of costs and benefits for those effecting or contemplating acts of violence.[56]

Other analysts, such as Jeffrey Herbst, have underscored the partisan nature of policy against conflict financing. He argued that sanction regimes are 'an implicit call for the military defeat of the rebels by the government', and that 'the vocabulary of victory and defeat has been transferred to the more neutral and technocratic language of sanctions and restraints on the trade of natural resources'.[57]

Given the changing strategic landscape, the quest for solutions to violent conflict will remain important in the years ahead – as will the search for a better understanding of conflict dynamics. A political-economy perspective avoids easy solutions based on single-factor explanations for violent conflict, and cuts through layers of misleading discourse about the nature and capabilities of armed groups. It also emphasises the global connectedness of conflict zones. The perspective focuses on the role of non-state armed groups in conflict dynamics, as well as the functions associated with violent conflict in the domestic institutional landscapes of states at war, and the private entities with which they are linked. The overt or clandestine roles of state security forces, security firms and other international actors in conflict zones create important domestic institutional dynamics that are geographically distant from battlegrounds, yet at the same time closely connected with them. A clear mapping of the distribution of costs and benefits from violent conflict across a global web of 'winners' and 'losers' may provide new entry points for policy, and may generate new contests in domestic politics.

Achim Wennmann is Senior Researcher at the Centre on Conflict, Development and Peacebuilding at the Graduate Institute of International and Development Studies, in Geneva, Switzerland.

Notes

1 This essay conceives of 'armed groups' as 'coherent, autonomous, non-state actors that rely on the threat or use of force to achieve their objectives'. Peter G. Thompson, *Armed Groups: The 21st Century Threat* (London: Rowman & Littlefield, 2014), p. 4. Yet this essay also acknowledges that the 'non-state' character of armed groups is frequently blurred by state sponsorship and related agendas.

2 The endnotes of this essay reference a small sample of the key works on the subject. For reviews, see Macartan Humphreys, *Economics and Conflict* (Boston, MA: Harvard University Press, 2002); Cynthia J. Arnson, 'The Political Economy of War: Situating the Debate', in Cynthia J. Arnson and I. William Zartman (eds), *Rethinking the Economics of War: The Intersection of Need, Creed, and Greed* (Baltimore, MD: Johns Hopkins University Press, 2005); and Achim Wennmann, 'Conflict Economies', in Vincent Chetail (ed.), *Post-Conflict Peacebuilding: A Lexicon* (Oxford: Oxford University Press, 2009), pp. 74–91.

3 Statement by Peter Maurer, 31 October 2015, https://www.icrc.org/en/document/peter-maurer-respect-laws-of-war.

4 *Ibid.*

5 See Brian Ganson and Achim Wennmann, *Business and Conflict in Fragile States: The Case for Pragmatic Action* (London: Routledge for IISS, forthcoming 2016).

6 US Department of Defense, 'Response to Congressional Inquiry on National Security Implications of Climate Related Risks and a Changing Climate', 23 July 2015, http://archive.defense.gov/pubs/150724-congressional-report-on-national-implications-of-climate-change.pdf?source=govdelivery.

7 Diane Davies, *Urban Resilience in Situations of Chronic Violence* (Washington DC and Cambridge, MA: USAID and MIT Centre for International Studies, 2012); Oliver Jütersonke, Robert Muggah and Dennis Rodgers, 'Gangs, Urban Violence, and Security Interventions in Central America', *Security Dialogue*, vol. 40, no. 4–5, 2009, pp. 373–97.

8 Carla Koppel and Anita Shama, *Preventing the Next Wave of Conflict: Understanding Non-Traditional Threats to Global Stability* (Washington DC: Woodrow Wilson International Centre for Scholars, 2003); Commission on Global Governance, *Our Global Neighbourhood* (Oxford: Oxford University Press, 1995); World Commission on Environment and Development, *Our Common Future* (Oxford: Oxford University Press, 1987).

9 Malcom Gladwell, *The Tipping Point: How Little Things Can Make a Big Difference* (New York: Back Bay Books, 2000), p. 9.

10 Office for the Coordination of Humanitarian Affairs (OCHA), *OCHA and Slow-Onset Emergencies* (New York: OCHA, 2011), p. 3.

11 Mary Kaldor, *New and Old Wars: Organised Violence in a Global Era* (Cambridge: Polity Press, 1999); Mark Duffield, *Global Governance and the New Wars: The Merging of Development and Security* (London: Zed Books, 2001).

12 Mats Berdal, 'How "New" are "New Wars"? Global Economic Change and the Study of Civil War', *Global Governance*, vol. 9, no. 4, 2003, pp. 477–502.

13 David Keen, *The Economic Functions of Civil War, Adelphi* 320 (London: Oxford University Press for the IISS, 1998), p. 11.

14 Martin Van Creveld, *The Transformation of War* (New York: Free Press, 1991), pp. 57–62.

15 Barry Buzan, *People, States and Fear: An Agenda for International Security Studies in the Post- Cold War Era* (London: Harvester Wheatsheaf, 1991).

16 See United Nations Security Council, 'Report of the Panel of Experts on Violations of Security Council Sanctions against UNITA', S/2000/203, 2000.

17 See Global Witness, *All the President's Men: The Devastating Story of Oil and Banking in Angola's Privatised War* (London: Global Witness, 2001); Global Witness, *Taylor Made: The Pivotal Role of Liberia's Forest in Regional Conflict* (London: Global Witness, 2001); and Global Witness, *A Rough Trade: The Role of Companies and Governments in the Angola Conflict* (London: Global Witness, 1998).

18 Ian Smillie, Lansana Gberie and Ralph Hazleton, *The Heart of the Matter: Sierra Leone, Diamonds and Human Security* (Ottawa: Partnership Africa Canada, 2000); Philippe Le Billon, 'The Political Ecology of War: Natural Resources and Armed Conflict', *Political Geography*, vol. 20, no. 5, 2001; Michael L. Ross, 'What Do We Know about Natural Resources and Civil War?', *Journal of Peace Research*, vol. 41, no. 3, 2004.

19 Jeffrey Herbst, 'African Militaries and Rebellion: The Political Economy of Threat and Combat Effectiveness', *Journal of Peace Research*, vol. 41, no. 3, 2004, p. 367.

20 Michael Pugh and Neil Cooper, *War Economies in a Regional Context: Challenges and Transformations* (Boulder, CO: Lynne Rienner, 2004), p. 97.

21 Paul Collier, Anke Hoeffler and Dominic Rohner, *Beyond Greed and Grievance: Feasibility and Civil War*, Centre for the Study of African Economies, 2007, p. 21.

22 Chritsopher Cramer, 'Homo Economicus Goes to War: Methodological Individualism, Rational Choice and the Political Economy of War', *World Development*, vol. 11, no. 11, 2002.

23 I. William Zartman, 'Need, Creed, and Greed in Intrastate Conflict', in Arnson and Zartman (eds), *Rethinking the Economics of War*, p. 257.

24 Karen Ballentine and Jake Sherman (eds), *The Political Economy of Armed Conflict: Beyond Greed & Grievance* (Boulder, CO: Lynne Rienner, 2003); Arnson and Zartman (eds), *Rethinking the Economics of War*.

25 Karen Ballentine, 'Beyond Greed and Grievance: Reconsidering the Economic Dynamics of Armed Conflict', in Ballentine and Sherman (eds), *The Political Economy of Armed Conflict*, pp. 259–60.

26 Ken Menkhaus, 'Governance without Government in Somalia: Spoilers, State Building, and the Politics of Coping', *International Security*, vol. 31, no. 3, 2007, p. 78.

27 David Keen, 'A Rational Kind of Madness', *Oxford Development Studies*, vol. 25, no. 1, 1997; David Keen, *The Economic Functions of Civil War* (London: IISS, 1998), p. 12.

28 *Ibid.*

29 Stathis N. Kalyvas, *The Logic of Violence in Civil War* (Cambridge: Cambridge University Press, 2006); Patricia Justino, Tilman Brück and Philip Verwimp (eds), *A Micro-Level Perspective on the Dynamics of Conflict, Violence, and Development* (Oxford: Oxford University Press, 2014).

30 Keen, 'A Rational Kind of Madness', p. 68; Christopher Cramer, *Civil War is Not a Stupid Thing: Accounting for Violence in Developing Countries* (London: C. Hurst & Co., 2006).

31 Rukmini Callimachi, 'Paying Ransoms, Europe Bankrolls Qaeda Terror', *New York Times*, 29 July 2014, http://www.nytimes.com/2014/07/30/world/africa/ransoming-citizens-europe-becomes-al-qaedas-patron.html.

32 *Ibid.*

33 *Ibid.*

34 James D. Fearon and David D. Laitin, 'Ethnicity, Insurgency and Civil War', *American Political Science Review*, vol. 91, no. 1, 2003, p. 88.

35 Radio Television Suisse, '"Le CICR Doit Établir une Relation avec le Groupe Etat Islamique"', 6 December 2015, http://www.rts.ch/info/monde/7312684-le-cicr-doit-etablir-une-relation-avec-le-groupe-etat-islamique.html.

36 Shiv Malik, 'The Isis Papers: Leaked Documents Show How Isis is Building its State', *Guardian*, 7 December 2015, http://www.theguardian.com/world/2015/dec/07/leaked-isis-document-reveals-plan-building-state-syria.

37 *Ibid.*

38 United States Institute of Peace, Glossary of Terms for Conflict Management and Peacebuilding, available at http://glossary.usip.org/resource/conflict-entrepreneur.

39 Mark Duffield, 'Globalisation, Transborder Trade, and War Economies', in Mats Berdal and David M. Malone (eds), *Greed*

and Grievance: Economic Agendas in Civil Wars (Boulder, CO: Lynne Rienner, 2000); R. Thomas Naylor, *Wages of Crime: Black Markets, Illegal Finance, and the Underworld Economy* (Ithaca, NY: Cornell University Press, 2002).

40 Pratap Chatterjee, 'Our Drone War Burnout', *New York Times*, 14 July 2015, http://www.nytimes.com/2015/07/14/opinion/our-drone-war-burnout.html.

41 Scott Shane, 'Drone Strikes Reveal Uncomfortable Truth: U.S. Is Often Unsure about Who Will Die', *New York Times*, 23 April 2015, http://www.nytimes.com/2015/04/24/world/asia/drone-strikes-reveal-uncomfortable-truth-us-is-often-unsure-about-who-will-die.html.

42 Chatterjee, 'Our Drone War Burnout'.

43 See Borderland Beat, 'Kingpin Strategy', 28 October 2012, http://www.borderlandbeat.com/2012/10/kingpin-strategy.html.

44 Patrick Corcoran, 'Mexico Has 80 Drug Cartels: Attorney General', InsightCrime, 20 December 2012, http://www.insightcrime.org/news-analysis/mexico-has-80-drug-cartels-attorney-general.

45 Private communication with the author.

46 Achim Wennmann, 'The Political Economy of Conflict Financing: A Comprehensive Approach Beyond Natural Resources', *Global Governance*, vol. 13, no. 3, 2007.

47 Achim Wennmann, *Conflict Financing and the Recurrence of Intra-state Conflict*, doctoral dissertation, Graduate Institute of International Studies and University of Geneva, 2007, pp. 203–4.

48 Rohan Gunaratna, ' Sri Lanka: Feeding the Tamil Tigers', in Ballentine and Sherman (eds), *The Political Economy of Armed Conflict*, p. 210.

49 Matthew Rosenberg, Nicholas Kulish and Steven Lee Myers, 'Predatory Islamic State Wrings Money from Those It Rules', *New York Times*, 29 November 2015.

50 Shiv Malik, 'The Isis Papers: Leaked Documents Show How Isis is Building its State'.

51 See 'IS Verdient Deutlich Weniger als Gedacht', Tagesschau, 6 November 2014, http://www.tagesschau.de/ausland/is-bnd-101.html.

52 *Ibid.*

53 Shiv Malik, 'The Isis Papers: Behind the "Death Cult" Image Lies a Methodological Bureaucracy', *Guardian*, 7 December 2015.

54 Achim Wennmann, 'Grasping the Financing and Mobilisation Cost of Armed Groups: A New Perspective on Conflict Dynamics', *Contemporary Security Policy*, vol. 30, no. 2, 2009.

55 'Islamic State: UN Resolution Aims to Curb Funding', BBC, 18 December 2015, http://www.bbc.com/news/world-middle-east-35120441.

56 Keen, 'A Rational Kind of Madness', p. 73.

57 Jeffrey Herbst, 'Economic Incentives, Natural Resources and Conflict in Africa', *Journal of African Economies*, vol. 9, no. 3, 2000, pp. 271.

Measuring the Threat from Foreign Fighters

Daniel Byman

On 13 November 2015, teams of attackers linked to jihadist group the Islamic State (also known as ISIS or ISIL) killed 130 people in Paris. Several of the attackers, including the mastermind, had fought in Syria, and their operation appeared to realise the fears of counter-terrorism officials in the United States and Europe: that the Syrian war would prove an incubator of terrorism on the scale of pre-9/11 Afghanistan. Yet not every war in the Muslim world that draws Western recruits produces international terrorism when these foreign fighters return home. Indeed, some, such as the conflicts in Somalia and Mali, have led to little terrorism 'bleedout'.[1]

For Western security officials, the risk seems overwhelming. FBI Director James Comey stated that 'there's going to be a diaspora out of Syria at some point and we are determined not to let lines be drawn from Syria today to a future 9/11'.[2] In part in response to this potential risk, in June 2014, the US once again intervened in Iraq, striking ISIS forces there and in Syria. President Barack Obama warned that, 'if left unchecked, these terrorists could pose a growing threat beyond that region, including to the United States.'[3]

As the Paris attacks demonstrate, these fears are not unfounded – but they can be overstated. The *potential* threat is tremendous, but the *actual* threat is lower. Many important jihads, such as the fight against the US in Iraq after 2003, produced little terrorism in the West. Effective policy can reduce the danger, preventing potential fighters from going to the war zone and disrupting their activities after they return. Nonetheless, there remains uncertainty about the terrorist threat posed by foreign fighters; why so many of these fighters decide not to join terrorist groups; and how effective policy can decrease the threat.

The threat of foreign fighters

Foreign-fighter returnees have played a role in various post-9/11 attacks, including the bombings in Madrid in 2004 and London in 2005, and have been linked to plots to carry out even larger assaults, such as the 2006 attempt to down more than ten transatlantic airliners.[4]

Analysis conducted by a leading scholar on foreign fighters, Thomas Hegghammer, identifies the dangers posed by returnees.[5] He observes that foreign-fighter veterans were involved in 107 of the 401 terrorist attacks or plots (see caveats below) instigated between 1990 and 2010, but that most foreign fighters do not become international terrorists. At most, one in nine foreign fighters return to attack the West; the real figure may be closer to one in 20. Hegghammer's findings indicate that returnees are more effective at carrying out operations than those who have not fought abroad. Veterans' involvement in plots increases the probability of success by a factor of around 1.5, and more than doubles the chances that the resulting attack will kill people. The majority of al-Qaeda operatives began as foreign fighters, and 'most transnational jihadi groups today are by-products of foreign fighter mobilizations'.[6]

Table 1: **Mobilisation of Foreign Fighters[7]**

Conflict Location	Duration	Estimated Number of Foreign Fighters
Afghanistan (anti-Soviet)	1980–92	5,000–20,000
Afghanistan (post-Soviet withdrawal)	1996–2001	1,000–1,500
Afghanistan–Pakistan (post-9/11)	2001–	1,000–1,500
Bosnia/Balkans	1992–95	1,000–2,000
Chechnya	1994–96	200–300
Yemen[8]	2000–	10
Iraq/Syria	2003–10 2010–	4,000–5,000 27,000–31,000
Somalia	2006–	200–400
Mali[9]	2012–	100–1,000

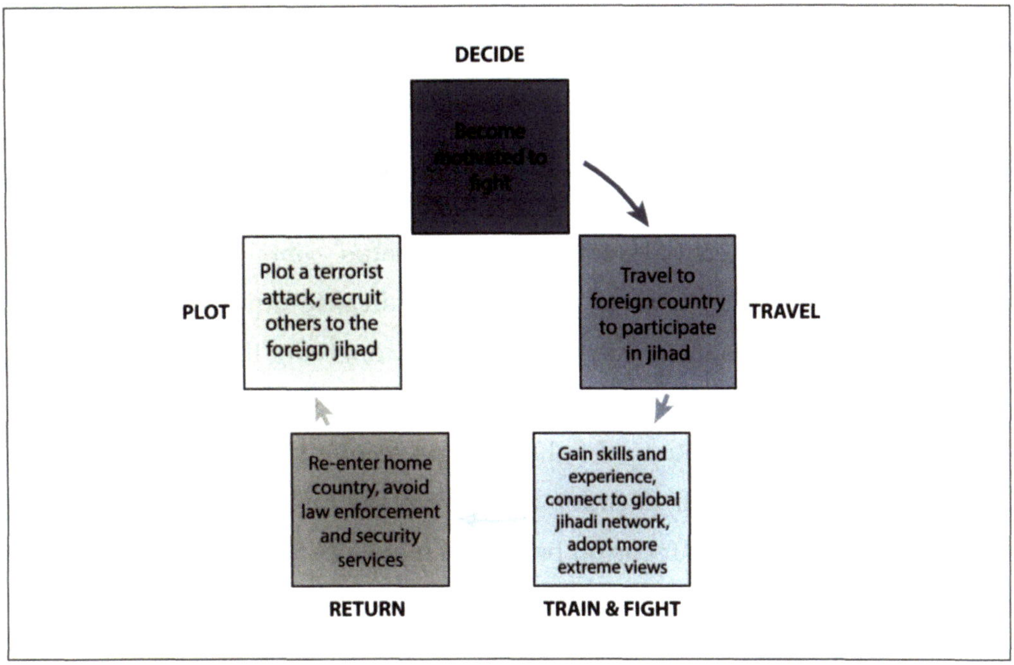

Figure 1: **Standard Schematic Model of Foreign-Fighter Radicalisation**

The process by which foreign fighters become returnees who pose a risk of terrorism can be framed in a rough, five-step model (see Figure 1).[10] The first step is for fighters to decide to join a conflict in a foreign country. At the start, the motivation for the fight abroad is usually specific and local – a desire to battle an invader or oppressor – rather than tied to a struggle against injustices at home. Overall, the decision to go fight in a foreign conflict is usually less an act of religious commitment than of rebellion and thirst for adventure. Indeed, two British foreign fighters on their way to Syria found it necessary to purchase *Islam for Dummies* and *The Koran for Dummies* for the trip.[11]

In the second step, the foreign fighters travel to the war zone. The aspiring jihadist makes contact with recruitment networks (or vice versa) and then begins the sometimes long and arduous journey to the foreign field of battle.

In the third step, the journey transforms as the recruits train and fight in the field of jihad. Training camps give recruits a sense of mission and focus, helping them develop *esprit de corps*.[12] While in the war zone, callow youths become seasoned combatants. As explained by MI5, Britain's domestic-security service, 'foreign fighters can gain combat experience, access to training and a network of overseas extremist contacts. The skills, contacts and status acquired overseas can make these individuals a much greater threat.'[13]

In the fourth step, the fighters return to their home countries in the West. Returned foreign fighters have enhanced their status – obtained a sort of 'street cred' – that makes them able to recruit and radicalise others.[14] Sympathisers often admire returnees from Iraq and Afghanistan, giving them the prestige to convince others to enter the fray.[15]

In the final step, the returned fighters begin to plot terrorist attacks in their homeland. They turn their attention away from the foreign struggle that took them abroad and begin to focus on perceived injustices at home: insults against Islam from their country's politicians and opinion leaders; support by their fellow countrymen or governments for forces oppressing Muslims abroad; and the daily prejudices faced by Muslims in the West, especially Europe.

As in past conflicts, foreign fighters going to wars in Iraq and Syria appear to be gaining valuable skills and access to terrorist networks.[16] Similarly, they are building networks with dangerous terrorist groups such as ISIS and Jabhat al-Nusra, the official al-Qaeda affiliate in Syria.[17]

Many of the concerns regarding Syria stem from the large number of foreigners who have fought or are fighting there. By the end of 2015, ISIS had attracted about 30,000 foreign Muslim fighters – most are from Muslim-majority countries, but around 5,000 are from the West.[18] Terrorism expert Peter Neumann contends that 'more people from Europe are being mobilized than in all the other foreign conflicts that have happened for the past 20 years taken together'.[19]

Ease of travel partly explains the large numbers. After the Syrian civil war began, in 2011, many volunteers simply flew, drove or took a train to Turkey and then slipped over the border into Syria. In addition, support for Syrian fighters is high among elements of Muslim communities in the West, particularly in Europe, according to Western intelligence officials, enhancing the status of volunteers within their communities back home.[20] And local Islamist groups play up the violence in war zones to gain recruits and funding. These local groups often link the conflict in Syria to problems at home, such as discrimination, or specific affronts such as cartoons mocking the Prophet Muhammad.[21] Part of the concern is that many fighters are apparently following the banner of ISIS – an almost cartoonishly evil group.[22] Advances in Iraq have given ISIS prestige and created the sense that it is a winner, making it more attractive to young recruits.

Furthermore, the emergence of Facebook, YouTube and Twitter in the 2000s provided jihadists with new tools to mobilise followers.[23] Social media has taken off around the world, and is playing a tremendously influential role in the Syria conflict. Groups such as ISIS are particularly skilled at using it to shape their image and recruit fighters.[24]

Is the risk of terrorism exaggerated?

A closer look at previous jihads indicates the limits of the five-step model, and why the threat – while still real – is likely to be less significant than security officials anticipate. Two sober, informed analysts warned in 2005 that the foreign-fighter flow from the war in Iraq could be 'even more dangerous than the fallout from Afghanistan' with regard to terrorism.[25] That same year, then-CIA director Porter Goss told the Senate Select Committee on Intelligence that 'Islamic extremists are exploiting the Iraqi conflict to recruit new anti-US jihadists.'[26] Security officials also warned that foreign fighters who went to other jihads – those in Somalia and Mali, for example – were a cause for concern.[27]

However, an analysis of major terrorist attacks in Europe between 2003 and 2014 reveals that Iraq and other conflicts did not play nearly as devastating a role as was feared. In fact, the number of individuals who went to fight in Iraq and came back to commit terrorist acts may actually be zero, and in any event is far lower than was originally anticipated.[28] Security analyst Teun van Dongen contends that, as 'plausible as this may have sounded, no successful terrorist attack has been carried out by a jihadist who gained experience by fighting in Iraq'.[29] Foreign fighters returning from conflicts in Somalia, Mali and elsewhere played no significant roles in major terrorist attacks. So, rather than the conflict in Syria mirroring those in Afghanistan in the 1980s and 1990s, it may be more reminiscent of the Iraq War in the 2000s.

Judgements on the danger created by foreign fighters are often skewed, partly because of confusion over how to measure the threat. The numbers cited by many terrorism experts should not (but do) include those already violent, radicalised and trained. Indeed, it is plausible that foreign jihads can actually serve as a diversion for those who have already been radicalised.[30]

Another measurement issue relates to the fact that some recruits – particularly those going to fight for ISIS after 2013 – were criminals who might have committed acts of violence anyway, although they would not have been considered terrorists if their crimes lacked a political context. Thus, if the measurement is violence generally, as opposed to political violence specifically, the impact of being a foreign fighter is more limited.

A final measurement issue relates to the conflation of plots with successful operations. When measuring *potential* danger, it is logical to include plots: had it not been for the security services disrupting the operation, people would have died. When measuring *actual* danger, however, a regular pattern of disrupted plots suggests the presence of competent counter-terrorism services (or incompetent terrorists).

Beyond these methodological caveats, there are many 'off-ramps' that prevent foreign fighters from returning and becoming successful terrorists (see Figure 2). Foreign fighters are by definition active in war zones, and many die; thus, in some instances, a foreign war reduces the danger at home by killing off would-be domestic terrorists. Although the numbers are vague, estimates by Western intelligence services suggest that around 20% of those who have travelled from the West to join the fight in Syria have died in combat, and that this percentage may be increasing due to internal disputes within militant groups.[31]

Some of those who survive do not return to their home country, because they are committed to finishing the fight abroad or they can no longer return without risking arrest. The motivation of perceived community defence, which led many to fight in the first place, applies more neatly to jihads in the Muslim world than

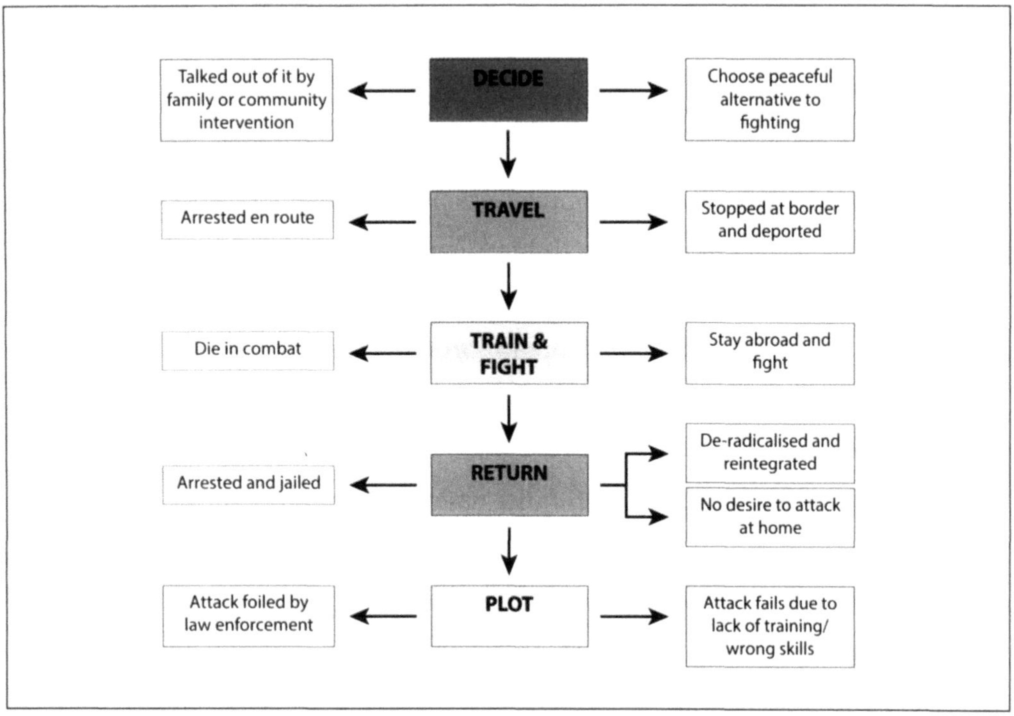

Figure 2: **Complex Model of Foreign-Fighter Radicalisation**

to terrorism in the West. In Syria and Iraq, many foreign fighters have made a point of declaring that they have no intention of returning home, and some have even posted videos of themselves burning their passports on the internet.[32]

Most foreign fighters do not die or choose to stay abroad, but they have several reasons for not taking up arms when they return. In some cases, the jihad itself may have discouraged them. Civil wars are brutal and often involve attacks on civilians, including fellow Muslims – particularly in the Syrian and Iraqi cases. Indeed, ISIS calls for killing supposed infidels and apostate Muslims, which it defines broadly. Infighting among jihadist groups is amplifying such disillusionment.

Some fighters may return glorifying jihad and proud of their role in the fighting, but that does not necessarily mean they want to bring the war home. As one study of foreign fighters in history found, 'the vast majority of them went on to lead essentially ordinary lives', even if they remained politically active.[33] While conflict abroad often enjoys legitimacy, terrorism at home does not. Thus, the social status that comes with fighting abroad is far greater than that of committing terrorist attacks at home. Finally, foreign fighters may not want to carry out attacks back in their home countries for the simple reason that such acts could endanger their friends and neighbours, either directly or through the resulting backlash. Nor are the groups these fighters join necessarily focused on the West. Organisations such as ISIS have anti-Western values and approve of attacks on the West – and, the November operation in Paris suggests, they are devoting some resources to this objective. However, as Hegghammer contends, the groups in Syria and Iraq today differ from al-Qaeda in Afghanistan in that they 'are not systematically trying to mount operations in the west',[34] and are far more focused on killing their Shia or Sunni enemies elsewhere.

Finally, when extrapolating from Afghanistan to the threat today, security services often miss the most important variable that determines the probability of attacks: the security services themselves. Before 9/11, only a small number of intelligence analysts worked on al-Qaeda, with even fewer focusing on the foreign-fighter issue. The *Washington Post* reports that US intelligence budgets have more than doubled since 9/11, and that Washington spent more than US$17 billion on counter-terrorism intelligence in 2013.[35] European states also increased their spending on counter-terrorism after 9/11, and continued to do so during the global financial crisis, despite severe budgetary constraints.[36] After the Paris attacks, this spending is likely to grow even more.

When Western foreign fighters travel abroad, their absence is usually noted by the local security services. Because so many returnees are known, it is hard for them to form the types of large local networks that would be necessary to carry out a major attack without attracting the attention of the security services. Here, social media is a double-edged sword. By publicising their activities online – so important to bragging rights back home – individuals identify themselves to security services. In addition, social media and the associated analysis of 'friends' and 'followers' can enable intelligence services to rapidly uncover broader networks of influence and conspiracy, and to discover when foreign fighters might return home.

Policy implications

Effective policy can reduce the dangers created by returnees even more dramatically. In the decision stage, it makes sense to reduce the numbers of those going to the conflict zone in the first place by interfering with their plans. To achieve this, European governments have set up programmes to work with communities, develop counter-narratives and provide alternatives to would-be fighters.

In the second stage, travel, disrupting the transit route via Turkey is one of the most effective ways of reducing the threat posed by foreign fighters returning to Europe and the US. Doing so will primarily require better cooperation between Western governments and the Turkish authorities. The latter stepped up their border-security efforts in 2015, but they have not always given a high priority to stopping fighters who seek to overthrow Syrian President Bashar al-Assad – some of whom are jihadists – from reaching the battlefield.[37]

Security cooperation among Western security services is also essential to disrupting travel. European security services had opportunities to disrupt some of the Paris plotters, but information was not always shared.[38] Intelligence collected from the communications of foreign fighters, shared open-source monitoring and other information can prove vital to uncovering transnational terrorist networks.

It is harder to reach foreign fighters when they are training and fighting in Iraq and Syria, but even here there are subtle ways to reduce the danger (in addition to military efforts to kill ISIS members). Western security agencies can sow doubt in the minds of extremist leaders in Iraq and Syria about the true loyalties of Western volunteers. If jihadist organisations come to view foreigners as potential spies or corrupting influences, they might execute them, assign them to non-combat roles, test their allegiances by offering them the one-way ticket of suicide bombings or even avoid recruiting them altogether.

Western security services report that they usually know when fighters return to their home country, and that many of them return with doubts about their actions abroad. The authorities can triage returnees to identify those who deserve the most attention. Harsh steps are at times necessary, but they must be applied carefully or else they may backfire. For example, the certainty of facing imprisonment upon return may make an individual feel that he or she has nothing to lose by moving towards terrorism at home. As analyst Shiraz Maher put it, 'this makes the decision for you'.[39] Such coercive measures can also create 'suspect communities' in which radicalisation is more likely, and whose members are discouraged from working with the police and government in general.[40]

To disrupt foreign fighters in the fifth stage, plotting terrorist attacks, the security services need to remain focused on the returnee problem, and to have sufficient resources to do so. Some of the jihadists who conducted attacks in France, Australia and elsewhere fell off the radar screen, despite being known to security services. This was partly due to resource constraints. So, ameliorating the foreign-fighter threat requires security services that are well funded.

As Figure 2 makes clear, there is no single factor or moment that prevents foreign fighters from conducting terrorist attacks back home. However, at each stage in the process, mitigating factors and policy interventions can reduce the likelihood of an attack. It is likely that some attacks will take place regardless. Yet it is important not to panic, and to recognise that both the US and Europe have dealt with this problem before, and already have measures in place to greatly reduce the threat of terrorism from jihadist returnees, as well as to limit the scale of any attacks that might occur.

Daniel Byman is a Professor in the Security Studies Program in the Edmund A. Walsh School of Foreign Service, and a Senior Fellow and Director of Research at the Center for Middle East Policy at the Brookings Institution.

Notes

1 This essay draws on my work with Jeremy Shapiro. See Daniel Byman and Jeremy Shapiro, 'Be Afraid. Be A Little Afraid: The Threat of Terrorism from Western Foreign Fighters in Iraq and Syria', Brookings Institution, November 2014, http://www.brookings.edu/~/media/research/files/papers/2014/11/western-foreign-fighters-in-syria-and-iraq-byman-shapiro/be-afraid--web.pdf; and Daniel Byman and Jeremy

Shapiro, 'Homeward Bound? Don't Hype the Threat of Returning Jihadists', *Foreign Affairs*, vol. 93, no. 6, November–December 2014, https://www.foreignaffairs.com/articles/iraq/2014-10-01/homeward-bound. I have expanded on this work in Daniel Byman, 'The Homecomings: What Happens When Arab Foreign Fighters in Iraq and Syria Return?', *Studies in Conflict & Terrorism*, vol. 38, no. 8, May 2015, pp. 1–22; and Daniel Byman, 'The Jihadist Returnee Threat: Just How Dangerous?', *Political Science Quarterly*, forthcoming in 2016.

2 Sari Horowitz and Adam Goldman, ' FBI Director: Number of Americans Traveling to Fight in Syria Increasing', *Washington Post*, 2 May 2014, http://www.washingtonpost.com/world/national-security/fbi-director-number-of-americans-traveling-to-fight-in-syria-increasing/2014/05/02/6fa3d84e-d222-11e3-937f-d3026234b51c_story.html.

3 White House, 'Statement by the President on ISIL', 10 September 2014, http://www.whitehouse.gov/the-press-office/2014/09/10/statement-president-isil-1.

4 Jeanine de Roy van Zuijdewijn, 'The Foreign Fighters' Threat: What History Can (not) Tell Us', *Perspectives on Terrorism*, vol. 8, no. 5, 2014, http://www.terrorismanalysts.com/pt/index.php/pot/article /view/378/html.

5 For Hegghammer's data on jihadist plots in general and on foreign fighters in particular, see Thomas Hegghammer, http://hegghammer.com/text.cfm?path=2176. For Hegghammer's most prominent research on foreign fighters, see Thomas Hegghammer, 'Should I Stay or Should I Go? Explaining Variation in Western Jihadists' Choice between Domestic and Foreign Fighting', *American Political Science Review*, vol. 107, no. 1, February 2013, pp. 1–14; and Thomas Hegghammer, 'The Rise of Muslim Foreign Fighters: Islam and the Globalization of Jihad', *International Security*, vol. 35, no. 3, pp. 53–94. Hegghammer used a very conservative estimate for outgoing foreign fighters. As a result, when he estimates the terrorism return rate at the end of the paper, that conservative estimate becomes an inaccurately low denominator. E-mail correspondence with Thomas Hegghammer, 22 April 2015.

6 Hegghammer, 'The Rise of Muslim Foreign Fighters', p. 53.

7 With the exception of the conflicts in Yemen, Iraq, Syria and Mali, this data is current as of 2010, according to Thomas Hegghammer. See Thomas Hegghammer, 'The Rise of Muslim Foreign Fighters: Islam and the Globalization of Jihad', *International Security*, vol. 35, no. 3, Winter 2010–11, pp. 53–94, http://belfercenter.ksg.harvard.edu/files/The_Rise_of_Muslim_Foreign_Fighters.pdf; and Soufan Group, 'Foreign Fighters: An Updated Assessment of the Flow of Foreign Fighters into Syria and Iraq', December 2015, http://soufangroup.com/wp-content/uploads/2015/12/TSG_ForeignFightersUpdate3.pdf.

8 Hegghammer, 'Should I Stay or Should I Go?', p. 5. The Yemen numbers may be inaccurately low, as reporting is limited.

9 David Lewis, 'Analysis: Mali's Islamist Groups United by War Threat', Reuters, 15 January 2013, http://www.reuters.

com/article/us-mali-rebels-analysis-idUSBRE90E0WQ20130115.

[10] For an implicit overview of this model, see Clint Watts, 'Countering Terrorism from the Second Foreign Fighter Glut', *Small Wars Journal*, May 2009, http://smallwarsjournal.com/jrnl/art/countering-terrorism-from-the-second-foreign-fighter-glut.

[11] Martin Robinson, ' British Terrorists from Birmingham Bought "Islam for Dummies" Book before Traveling to Syria to Join Rebel Fighters in Jihad', *Daily Mail*, 8 July 2014, http://www.dailymail.co.uk/news/article-2684714/I-tell-I-m-going-jihad-Lol-I-ll-arrested-What- British-terrorist-Birmingham-told-childhood-friend-travelled- Syria-join-rebel-fighters.html.

[12] Interview with Peter Neumann, 17 June 2014.

[13] MI5, 'Foreign Fighters', https://www.mi5.gov.uk/home/about-us/what-we-do/the-threats/terrorism/international-terrorism/foreign-fighters.html.

[14] Center for Terroranalyse, 'The Threat to Denmark from Foreign Fighters in Syria', 4 December 2014, p. 5, https://www.pet.dk/English/CenterforTerrorAnalysisCTA/~/media/CTA/20131203SyrienUKLASSENGFINALpdf.ashx.

[15] For examples of this argument, see Marc Sageman, *Understanding Terror Networks* (Philadelphia: University of Pennsylvania Press, 2004); and Evan F. Kohlmann, *Al-Qaida's Jihad in Europe: The Afghan–Bosnian Network* (New York: Berg, 2004). See also Hegghammer, 'Should I Stay or Should I Go?', p. 12.

[16] Center for Terroranalyse, 'The Threat to Denmark from Foreign Fighters in Syria', p. 1.

[17] Meir Amit Intelligence and Terrorism Information Center, 'Foreign Fighters in Syria', December 2013, p. 22, http://www.terrorism-info.org.il/en/article/20607.

[18] Soufan Group, 'Foreign Fighters', p. 6. For a study of the flow of Shia fighters to Syria, see Phillip Smyth, 'The Shiite Jihad in Syria and Its Regional Effects', *Washington Institute*, February 2015, http://www.washingtoninstitute.org/policy-analysis/view/the-shiite-jihad-in-syria-and-its-regional-effects; Nicholas Rasmussen, 'Countering Violent Islamist Extremism: The Urgent Threat of Foreign Fighters and Homegrown Terror', remarks delivered before the US House Committee on Homeland Security, 11 February 2015, http://www.nctc.gov/docs/Countering_Violent_Islamist_Extremism.pdf.

[19] Ella Flaye, 'At Least 500 Europeans Fighting with Syria Rebels, Study Finds, Stoking Radicalization Fears', CBS, 30 April 2013, http://www.cbsnews.com/8301-202_162-57582040/at-least-500-europeans-fighting-with-syria-rebels-study-finds-stoking-radicalization-fears/.

[20] Interview with European government official, June 2014.

[21] Shaarik H. Zafar, 'Western Foreign Fighters in Syria: Implications for US CVE Efforts', remarks delivered at the Washington Institute for Near East Policy, 14 March 2014, http://www.washingtoninstitute.org/uploads/Documents/other/ZafarPreparedRemarks20140314.pdf.

[22] Daniel Byman, *Al Qaeda, the Islamic State, and the Global Jihadist Movement* (Oxford: Oxford University Press, 2015), pp. 163–86; William

McCants, *The ISIS Apocalypse: The History, Strategy, and Doomsday Vision of the Islamic State* (New York: St. Martin's Press, 2015).

23 Shiraz Maher, 'ICSR Insight: British Foreign Fighters in Syria', International Centre for the Study of Radicialisation and Political Violence, 15 October 2013, http://icsr.info/2013/10/british-foreign-fighters-in-syria/.

24 Jeff Stone, ' ISIS Attacks Twitter Streams, Hacks Accounts to Make Jihadi Messages Go Viral', *International Business Times*, 17 June 2014, http://www.ibtimes.com/isis-attacks-twitter-streams-hacks-accounts-make-jihadi-message-go-viral-1603842.

25 Peter Bergen and Alec Reynolds, 'Blowback Revisited', *Foreign Affairs*, vol. 84, no. 6, November–December 2005, http://www.foreignaffairs.com/articles/61190/peter-bergen-and-alec-reynolds/blowback-revisited.

26 Porter Goss, 'DCI Testimony: Global Intelligence Challenges 2005', remarks delivered before the US Senate Select Committee on Intelligence, 16 February 2005, https://www.cia.gov/news-information/speeches-testimony/2005/Goss_testimony_02162005.html.

27 'MI5 Chief Warns of Terror Threat from Britons Trained in Somalia', *Guardian*, 17 September 2010, http://www.theguardian.com/politics/2010/sep/17/mi5-chief-somalia-terro-threat; Elaine Ganley, 'France Girds for New Threats amid Mali Operation', Associated Press, 15 January 2013, http://bigstory.ap.org/article/france-girds-new-threats-after-mali-operation. For statements on the seriousness of the Somali threat, see Andrea Elliott, 'A Call to Jihad Answered in America', *New York Times*, 11 July 2009, http://www.nytimes.com/2009/07/12/us/12somalis.html.

28 Several cases involve indirect links to Iraq, such as the bungled June 2007 attack in Glasgow and the 2010 attack in Stockholm.

29 Teun van Dongen, 'Foreign Fighters in Syria: A Terrorist Threat to Europe?', Aspen Institute Italia, 11 April 2013, paragraph 7, https://www.aspeninstitute.it/aspenia-online/article/foreign-fighters-syria-terrorist-threat-europe.

30 Hegghammer, 'Should I Stay or Should I Go?', p. 6.

31 Interview with a European Intelligence officer, May 2014.

32 See 'European ISIS Foreign Fighters (Jihadist) Burn Their Passports after Arriving in Syria', LiveLeak, 27 June 2014, http://www.liveleak.com/view?i=509_1403900374.

33 David Malet, 'Foreign Fighters' Playbook', *Foreign Affairs*, 8 April 2014, https://www.foreignaffairs.com/articles/africa/2014-04-08/foreign-fighters-playbook.

34 Thomas Hegghammer, 'Counter-terrorism', remarks delivered before the UK House of Commons Home Affairs Committee, 11 February 2014, http://www.publications.parliament.uk/pa/cm201314/cmselect/cmhaff/231/231.pdf.

35 Wilson Andrews and Todd Lindeman, 'The Black Budget', *Washington Post*, 29 August 2013, http://www.washingtonpost.com/wp-srv/special/national/black-budget/; Ewen MacAskill and Jonathan Watts, 'US Intelligence Spending Has Doubled Since 9/11, Top Secret Budget Reveals', *Guardian*,

29 August 2013, http://www.theguardian.com/world/2013/aug/29/us-intelligence-spending-double-9-11-secret-budget.

[36] See 'U.K. Increases Intelligence Agencies' Budget to Counter Terror Threat', *Homeland Security NewsWire*, 26 June 2013, http://www.homelandsecuritynewswire.com/dr20130626-u-k-increases-intelligence-agencies-budget-to-counter-terror-threat.

[37] 'Turkey to Boost Border Security in Wake of ISIL Attack', Al-Jazeera, 25 July 2015, http://www.aljazeera.com/news/2015/07/turkey-bolster-border-security-suruc-attack-150721000033031.html; Tim Arango and Eric Schmitt, 'A Path to ISIS, Through a Porous Turkish Border', *New York Times*, 9 March 2014, http://www.nytimes.com/2015/03/10/world/europe/despite-crackdown-path-to-join-isis-often-winds-through-porous-turkish-border.html.

[38] Jim Yardley et al., 'Inquiry Finds Mounting Proof of Syria Links to Paris Attacks', *New York Times*, 15 November 2015, http://www.nytimes.com/2015/11/16/world/europe/inquiry-finds-mounting-proof-of-syria-link-to-paris-attacks.html.

[39] Interview with Shiraz Maher and Peter Neumann, 17 June 2014.

[40] Floris Vermeulen, 'Suspect Communities – Targeting Violent Extremism at the Local Level: Policies of Engagement in Amsterdam, Berlin, and London', *Terrorism and Political Violence*, vol. 26, no. 2, 2014, pp. 286–306.

The Shifting Roles of Female Combatants

Carol Mann

Throughout 2015, the international media paid great attention to female combatants of all kinds: glamourised Kurdish fighters; those taking up arms for the Islamic State (also known as ISIS or ISIL), including suicide bombers and the group's all-female brigade, al-Khansaa; and Hayat Boumedienne and Hasna Ait Boulahcen, who worked alongside male jihadists in France in January and November respectively.

In many armed conflicts, women position themselves as modern-day warriors fighting for a political and/or religious cause against enemies they perceive to be excessively powerful and morally unacceptable. Even when working alone, as in the case of suicide bombers, female combatants tend to consider their actions as part of a wider military campaign: Palestinian or Chechen suicide bombers envisage their sacrifice as being their contribution to a wider, all-encompassing cause. This is the main factor separating them from rampage killers, who are usually male, and generally devoid of any ideological motivations.

Armed militancy among women appears to be on the rise. Stereotypes that create gendered discourses on war and peace disappear on battlefields expanded to include public places such as schools, theatres, hospitals, supermarkets and anywhere else civilians can be held hostage. In this context, militarised violence seems to be taking place in an increasingly de-gendered conflict space, as physical strength is not a prerequisite for modern forms of attack.

Yet in many war zones, especially those in the Middle East, traditional patriarchal structures remain in place and may recently have been strengthened, for a variety of reasons. The decision taken by female militants (as opposed to their male counterparts) to participate in conflict is usually considered antisocial and unacceptable within their societies, as it defies not only gender norms but also widely held notions of propriety.

Even in the West, the language that the media uses to describe female terrorists seeks invariably to diminish their political motivations, reducing it to some kind of – at best – psychotic impulse even more than for men. Although male ISIS fighters are often described as psychopaths killing for killing's sake, their actions are frequently legitimised by religious references brought into the discourse – which is not necessarily the case for their female counterparts – hence the typical tabloid story 'Skanky Suicide Bomber Used to Be a Selfie-taking Party Animal', a reference to Boulahcen, accompanied by countless photographs of her in the bath.[1] Her male comrades-in-arms have rarely been treated with such flippancy, but are usually considered responsible for their acts and therefore held to be truly dangerous, a feature regularly brought out by leading articles in the more serious French press.[2]

In the 1990s, the female entourage of Abimael Guzmán – then leader of Shining Path, a Peruvian Maoist insurgent group – were described as 'sex objects', among whom Maritza Garrido Lecca was a particularly lethal 'femme fatale'.[3] Female Kurdish combatants, fighting in both the Syrian People's Protection Units (YPG) and for Iraq's Kurdistan Regional Government (KRG), known here as Peshmerga,[4] have been the subject of many romanticised articles.[5] Such pieces often include photographs of the youngest and most alluring of these women, and one company even marketed an outfit the tabloid press termed 'Peshmerga chic', setting off a storm on social media. The real problem is that violence perpetrated by women, as opposed to that meted out to them as victims, remains unacceptable in general opinion and has only recently become the subject of scholarly research, starting with work on female Nazi concentration camp guards. For ISIS, women are allowed to fight only as a last resort; according to the al-Khansaa manifesto, they may participate in jihad 'if the enemy is attacking her country and the men are not enough to protect it and the imams give a fatwa for it, as the blessed women of Iraq and Chechnya did, with great sadness'.[6] In other words, the group demands that violent women be controlled by still more violent men.

Brief history of female combatants

There are also other circumstances in which it is socially acceptable for women to fight, for the shortest possible time and in a national emergency. Such behaviour is validated by the presence of a consensually defined enemy defined as such by the fighter's community, who is perceived as foreign/alien or an outsider (even when they live in the same place as the protagonist, as is often the case with ethnic and sectarian violence). Whatever the configuration, women remain visibly subordinate to a male chain of command and, more often than not, do not benefit from public recognition of their achievements. Thus, Joan of Arc, who became a national heroine in France by defending her country against an English invasion, was considered to be, in modern terms, an illegal female combatant – and she was treated as such. As a result, she was burned alive on the excuse of heresy for her acts of subversion by the English and their French collaborators. It was not until 1920 that she was canonised; the French government subsequently established a national festival in her honour, in an effort to reinvigorate national pride after the disasters of the First World War.

The female resistance fighter, half-way between regular soldier and guerrilla, is an important archetype in history. Despite its reluctance to allow female participation in public spaces, ISIS has repeatedly produced propaganda that refers to the female warriors of the Koran, such as Khadija Zeyneb, Aisha and Safiya bint Abdul Muttalib.[7] Although they are not freedom fighters, these women are regarded by ISIS as warriors united by a common cause, and as role models for would-be combatants.

During the Second World War, in Western Europe, female resistance often took the form of individual acts of heroism that aligned with the gender norms of the time: nursing or hiding male fighters; smuggling children out of war zones; transporting munitions and weapons; and occasionally participating in fighting. The Algerian women who fought against French colonists in the late 1950s and early 1960s fulfilled many of the same roles. Even today, women in rebel groups everywhere, including ISIS, are involved in such activities.

In almost every insurgency, the tasks performed by female militants are similar to peacetime undertakings, and are often as dangerous as being on the front-line, yet they lack the kudos of handling weapons afforded to men. In peace talks, women are frequently doomed to invisibility because of the participants' (including international authorities who broker such talks) disregard for non-military resistance.

Nonetheless, female militants are treated as enemies of state by the authorities, often receiving particularly brutal treatment after being captured. This has been the case with those who resisted the Nazis and various Latin American dictatorships, as well as, more recently, with Kurdistan Workers' Party (PKK) fighters caught by the Turkish police.[8]

To a certain extent, the Soviet Union's egalitarianism provided an alternative that Marxist–Leninist militant groups have looked to for inspiration and guidance: a structure in which women can participate in decision-making and combat (rather than one organised on purely feminist and pacifist principles). This was evident during the 1936–39 Spanish Civil War with the communist Republican *milicianas* active on the front-lines, and even more so with the anarchist feminist group Mujeres Libres (Free Women) – the nearest modern equivalent of which is perhaps the female branches of the PKK and the YPG.

Inspired by the achievements of female partisans during the Second World War, particularly in the Soviet Union, Marxist–Leninist revolutionary groups from the Vietcong and the Indian Naxalites to Colombia's FARC have included female combatants in their formations. In the early days, attaining equality in gender relations was, even for female recruits, a secondary matter to the struggle against an overwhelmingly powerful enemy – generally, a well-equipped army. The participation of women in militant and revolutionary organisations on a (more or less) equal footing occurred from the late 1960s onwards, coinciding with protests by Marxist-inspired students, workers and minority groups, and especially women's liberation movements.

In the 1970s, Maoism started to guide the activities of rebels across the globe, all of whom were inspired by the Vietcong's resistance to the US invasion of Vietnam. These ideologies provided the dominant progressive rhetoric and guidance to counter capitalism and imperialism, and in each rebel group female combatants had, with significant variations, their allotted spaces.

The fall of the Soviet Union and China's embrace of capitalism prompted a sharp decline in the material and ideological resources available to Marxist groups. Cuba still provides some support to such organisations in Latin America (which is why the negotiations between FARC and the Colombian government have been held in Havana). And North Korea has trained insurgent movements all over the world – including the Sri Lankan Janatha Vimukthi Peramuna, the Palestine Liberation Organisation and Western Sahara's Polisario Front – not to promote socialism and women's rights, but to fight a nebulous Western imperialism that includes its allies such as Morocco.

Recruitment in armed movements

Why do some women wage war in armed movements and place themselves in a position where they can only be considered enemies of the state? The notions of gender deviance and perversion (of which Joan of Arc was accused) are no longer deemed relevant by social scientists. Indeed, state institutions such as the armed forces and police in many countries are developing marketing strategies to recruit female personnel. To quote a site geared towards women maintained by the Australian Defence Force: 'right now, thousands of women of all ages and from every walk of life are enjoying rewarding careers in the Navy, Army and Police Force.'[9] Similar campaigns, complete with alluring videos, are being run in most countries seeking to enlist part of the female workforce, including those in which there is a real possibility of joining competing insurgencies, such as Colombia.[10]

Indeed, a number of governments have become increasingly aware that women may choose a military or paramilitary career for all kinds of personal and social reasons, including their experience growing up watching the same hyper-masculine TV shows, movies and video games as their brothers (not to mention female action heroes who have hyper-masculine values).

Ideology may drive middle-class women to join armed groups, particularly those such as the Italian Red Brigades, the German Red Army Faction and their social equivalents in Latin America. Such women have had more career opportunities than most – as can be seen in the case of Aqsa Mahmood, a former medical student from Glasgow who is now a high-ranking member of al-Khansaa in Syria.[11] Furthermore, migration to urban areas has allowed access to education for girls who might have otherwise barely been literate, especially in Latin America, and to a certain extent in Asia. The same could be said for the daughters of people who migrated from rural Pakistan to the United Kingdom, or from the Maghreb to France. From the 1970s onwards, high schools and universities have provided spaces for the mobilisation of young women eagerly discovering alternative lifestyles away from home. At the same time, such women have been able to identify and understand grievances that had hitherto been internalised. Radicalisation has sometimes provided them with a new language, and joining radical movements has appeared as an empowering means of escape from the drudgery many associate with women's traditional roles.

Some also have social and personal reasons for joining such groups – including poverty, social and political exclusion – and sometimes a desire for revenge. The influence of these factors is specific to the individual, and they may also extend to a need for redemption from gender-based suffering and sexual violence. The efforts of recruiters, ranging from mild persuasion to veiled coercion, play a significant part in the process. Thus, the act of joining is not always the product of careful consideration. Discussing activists in El Salvador, Jocelyn Viterna proposes three paths to membership of guerrilla movements that may well apply in other countries: politicised guerrillas, reluctant guerrillas and recruited guerrillas.[12] As argued by Simon Bonnet, the ways to jihad are complementary to one another: religious and political commitment interact with a feeling of belonging to a group and a sense of purpose.[13] The short interval between conversion to radical Islam and departure to Syria among European female recruits demonstrates this combined sense of urgency and anguish.

The situation reflects a significant trend: the presence of women in regular militaries is increasing in many parts of the world, a surge that may relate more to shrinking employment opportunities than to true military ambitions or patriotism (unlike in revolutionary movements). In some cases, the inclusion of former combatants in military units is an astute attempt at reconciliation. Thus, 26.6% of the South African National Defence Force consists of women (70% of whom are black), which may reflect their participation in liberation movements in the 1970s and 1980s.[14]

The choice between joining a regular army or illegal combat units is a vital, existential one. Legitimacy and legality are at loggerheads when it comes to action: violent acts carried out by the regular military remain legal and an intrinsic part of its mandate, whereas identical acts by a militant are criminal in the eyes of the law, regardless of how they are justified by her group. Yet problems facing female recruits in the military resemble those encountered by women who join an insurgency. Female fighters on both sides have often had difficulty asserting their authority over male soldiers and been pushed into auxiliary positions, with the exception of those who have joined FARC, the PKK and the YPG.[15]

The places that these women create for themselves in various movements are determined by significant local differences, even though the internal distribution of power usually concedes supreme authority to men. Yet the potential for upward mobility associated with women in armed groups is inspirational for potential recruits. Women who choose to be combatants take far greater risks than their male counterparts. Fighting may constitute a short phase in a man's life, which he may or may not capitalise on afterwards. By contrast, women may not recover from such a radical decision, professionally or emotionally, especially if they are neither recognised as combatants nor allowed to participate in peace talks. With the exception of male European jihadists, men, especially in South America, Asia and the Great Lakes region of East Africa, enjoy family support and leave survival issues to their wives and mothers at home; this is not the case for female militants, who often find themselves isolated, both as combatants and after demobilisation. Becoming a guerrilla may be a permissible career option for married men, but it rarely is for married women, except in movements in which their families are already involved.[16]

In research that begins with guerrilla movements in Latin America and extends to those in Asia, Margaret Gonzalez-Perez has shown that women have tended to be more active and powerful in revolutionary movements with a domestic agenda than in those with an international outlook (in the latter, the top leadership positions have remained in the hands of a male elite).[17] This may be because women have always had closer links to, and knowledge of, features of their immediate environments, including neighbours, schools and health facilities, whereas men have usually worked at some distance from home, including abroad as menial labourers or factory workers. In an age of rapid communication, this line of reasoning applies to not only knowledge of local rather than national issues, but also the domestic rather than international arena. Since the end of the Cold War, the majority of these groups have focused on local issues rather than changing the world order (al-Qaeda, ISIS and other Islamist-aligned groups are a notable exception to this).

Today, growing social injustice, marginalisation and feelings of hopelessness are proving powerful motivations for joining armed groups. This is certainly the case for Palestinian resistance fighters, including suicide bombers, who for the past 70 years have had few prospects of a positive future, and who seem to be motivated by sheer despair. After the fall of the Soviet Union and the ensuing ideological void, it is possible that the despair factor may increasingly supplant ideological motives.

This tendency also applies to the young women who join ISIS.[18] Their motivations often seem to come from grievances related to the difficulty of integrating into middle-class, liberal Western society, in which professional accomplishment is the only permissible path to social acceptance and status. The family and religious values that are denigrated by more trendy popular culture capitalising on some kind of potted feminism (TV series such as *Sex in the City*, *Girls* and *Gossip Girls*) are precisely those that are praised in ISIS rhetoric, which seeks to capitalise on deep-seated frustrations and anger. Marrying unknown combatants is one way of joining the jihad; the al-Khansaa brigade allows its members to be active in the public space as armed – sometimes brutal – policewomen who, like the morality police of Iran or Sudan, enforce religious law. Such publicly performed acts of violence are considered a form of jihad and sacrifice as they imply breaking religious law, namely consorting with *na mahram* (men to whom the policewomen are not related), which is theoretically forbidden in Islam. In southern Afghanistan, the government-sanctioned presence of women in positions of authority is perceived by Taliban and their allies to be an offence against wider society. This is why they have repeatedly targeted policewomen in Kandahar. As a younger and more media-savvy group, ISIS has found innovative, opportunistic ways to circumvent traditional prohibitions, such as a recent fatwa that allows the formation of a female battalion.[19]

Thus, concessions are woven into religiously cautioned social obligations in the name of an overarching moral ideal, as has always been the case in honour-based societies (in which so-called 'honour killings' often occur). Such compromises have also been made by the YPG, without any specifically religious motivations, to show respect for the morals and practices of local ethnic groups, in particular the parent generation of their female recruits going, on an equal basis, into what have been hitherto considered strictly male combat zones. In return for these concessions, female fighters are required to abide by a stringent moral code, which in the case of the YPG and the PKK means giving up their sexuality entirely. Their sworn chastity makes their dedication respectable in the eyes of their families, and validates their participation in combat for wider society. On both sides of the conflict between ISIS and the YPG, the female body is ultimately given up to the collective cause. Suicide bombing by women is considered by both groups to be the ultimate form of martyrdom: in Middle Eastern and Central Asian cultures, self-immolation is both an ancient tradition and a phenomenon on the increase for a variety of reasons.[20] This can be seen in the case of Arin Mirkan, who killed ten ISIS fighters in Kobane by detonating the explosives she was carrying. Posters and badges bearing her image are now ubiquitous in Turkish and Syrian Kurdistan, as this writer has been able to observe first-hand. Female warriors are also honoured by ISIS, which asked in January 2015 for the release of Sajida Mubarak al-Rishawi – an Iraqi woman arrested in Jordan in 2005, after she failed to carry out a suicide bombing – in exchange for the return of a kidnapped Jordanian pilot. However, ISIS appears to want to tightly control female jihadists, for fear that they may challenge the group's domination of women.

It is unclear whether the increasing presence of women on the front-lines of many conflicts, as part of both regular militaries and non-state armed groups, is a progressive, emancipatory trend caused by an expansion of battlefields and a blurring of traditional gender roles. Despite the constraints they face, women in regular armies, as well as groups such as the YPG, Shining Path and the Sri Lankan Liberation Tigers of Tamil Eelam, have gained a novel measure of empowerment and equality in male-dominated environments. Whether this will translate into greater rights and opportunities in wider society, especially after demobilisation, is quite another matter.

Carol Mann is a sociologist specialising in gender and armed conflict, and the director of Women in War (Paris).

Notes

1 Natalie Musumeci and Bruce Golding, 'Skanky Suicide Bomber Used to Be a Selfie-taking Party Animal', *New York Post*, 20 November 2015, http://nypost.com/2015/11/20/skanky-suicide-bomber-used-to-be-a-selfie-taking-party-animal/.

2 Emeline Cazi, 'La Fratrie Kouachi, de la Petite Délinquance au Djihad', *Le Monde*, 9 January 2015, http://abonnes.lemonde.fr/attaque-contre-charlie-hebdo/article/2015/01/08/attaque-a-charlie-hebdo-que-sait-on-des-deux-suspects-recherches_4551181_4550668.html.

3 Camille Boutron, 'Women at War, War on Women', in Seema Shekhawat (ed.), *Female Combatants in Conflict and Peace: Challenging Gender in Violence and Peace* (London: Palgrave Macmillan, 2015), p. 15.

4 A term erroneously ascribed to all Kurdish female combatants, both in Iraq and in Syria.

5 Women make up more than 40% of KRG forces and around 30% of the YPG (which

has close links with the Turkish Kurdistan Workers' Party, a left-wing Turkish militant organisation founded in 1978 by Abdullah Ocalan), although their participation in the latter group is decreasing. Female fighters are active on the front-lines in the battle against ISIS. Agri Ismail, 'Is Peshmerga Chic Offensive?', Al-Jazeera, 11 October 2014, http://www.aljazeera.com/indepth/opinion/2014/10/peshmerga-chic-offensive-2014101173639489234.html.

6 ISIS, 'Women of the Islamic State: A Manifesto on Women by the Al-Khanssaa Brigade', Charlie Winter (trans.), February 2015, https://www.quilliamfoundation.org/wp/wp-content/uploads/publications/free/women-of-the-islamic-state3.pdf.

7 Simon Bonnet, 'Western Women in Jihad, Triumph of Conservatism or Export of Sexual Revolution?', Diploweb, 12 June 2015, http://www.diploweb.com/Western-women-in-jihad-triumph-of.html.

8 As in the case of Kevser Elturk, who was killed by Turkish police. 'Picture of Naked, Bloodied "PKK Female Militant" Sparks Online Rage', Al-Arabiya, 19 August 2015, http://english.alarabiya.net/en/News/middle-east/2015/08/19/Picture-of-naked-bloodied-PKK-female-militant-sparks-online-rage-.html.

9 Australian Defence Force, 'Women in the ADF', http://www.defencejobs.gov.au/women/.

10 See videos posted online by the Colombian army, such as 'Formación de las Mujeres Suboficiales del Ejército Colombiano', https://www.youtube.com/watch?v=4Aw6hxoN7LM.

11 Soeren Kern, 'Britain's Female Jihadists', Gatestone Institute, 21 September 2014, http://www.gatestoneinstitute.org/4714/britain-female-jihadists.

12 Jocelyn S. Viterna, 'Women's Mobilization into the Salvadoran Guerilla Army', in Jeff Goodwin and James M. Jasper (eds), *The Social Movements Reader: Cases and Concepts* (Oxford: Wiley-Blackwell, 2014), p. 88.

13 Bonnet, 'Western Women in Jihad'.

14 'Fact File: Regular Force Levels by Race & Gender: April 30, 2011', defenceWeb, 20 June 2011, http://www.defenceweb.co.za/index.php?option=com_content&view=article&id=16708:fact-file-sandf-regular-force-levels-by-race-a-gender-april-30-2011-&catid=79:fact-files&Itemid=159.

15 Susanne Conze and Beate Fieseler, 'Soviet Women as Comrades-in-Arms: A Blind Spot in the History of the War', in Robert W. Thurston and Bernd Bonwetsch (eds), *The People's War: Responses to World War II in the Soviet Union* (Chicago, IL: University of Illinois Press, 2000), pp. 216–21.

16 One such example could be the Syriac Christian women's militia, which formed in December 2015 and includes married women. 'Syriac Christian Women Take up Kalashnikovs to Fight Islamic State', Agence France-Presse, 13 December 2015, http://www.theguardian.com/world/2015/dec/13/syriac-christian-women-militia-fight-islamic-state.

17 Margaret Gonzalez-Perez, 'From Freedom Birds to Water Buffaloes: Women Terrorists in Asia', in Cindy D. Ness (ed.), *Female Terrorism and Militancy: Agency, Utility, and*

Organization (Abingdon: Routledge, 2008), p. 183.

[18] Armelle Le Bras-Chopard, 'Femmes Françaises Djihadistes en Syrie et en Irak', paper presented at Women in War's annual conference, Beirut, June 2015, http://womeninwar.org/wordpress/wp-content/uploads/2015/08/Beirut/4/1.pdf.

[19] 'Les Limites de l'Opération Française Contre l'E.I.', *Le Monde*, 13–14 December 2015.

[20] Elizabeth Campbell and Isabelita Guiao, 'Muslim Culture and Female Self-Immolation: Implications for Global Women's Health Research and Practice', *Health Care for Women International*, vol. 25, no. 9, 2004.

Chapter Two

Maps, Graphics and Data

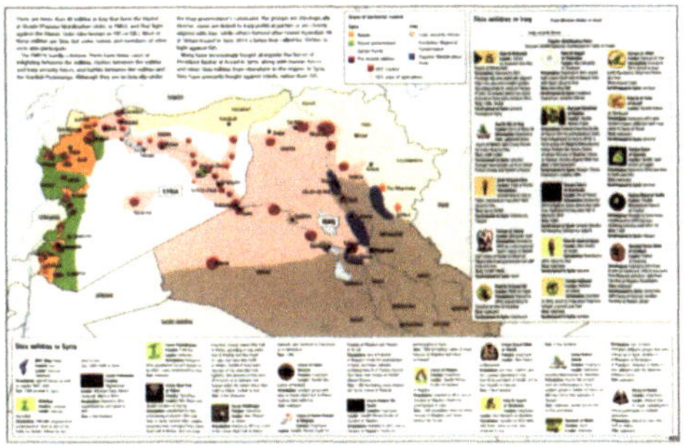

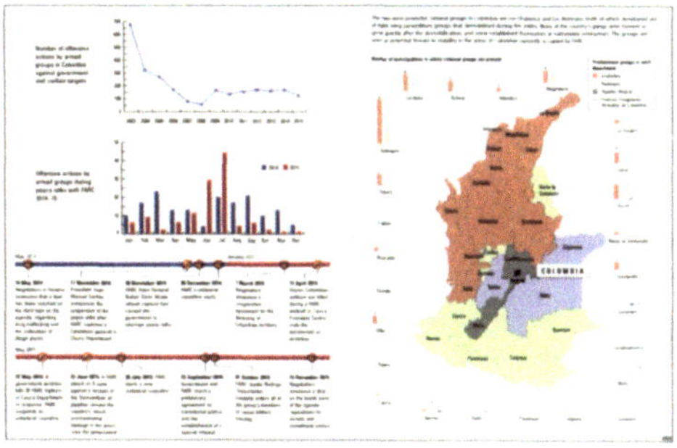

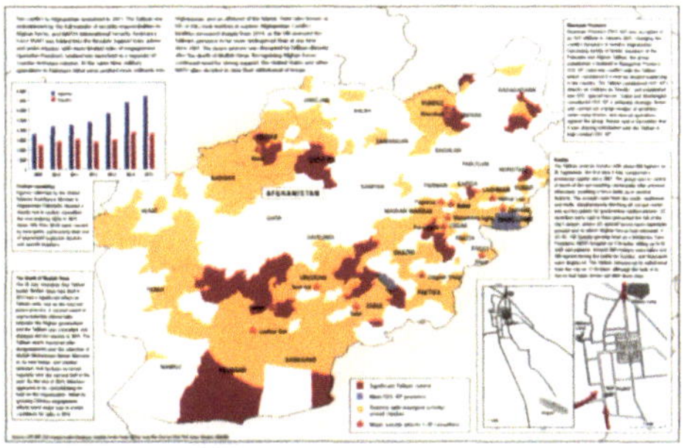

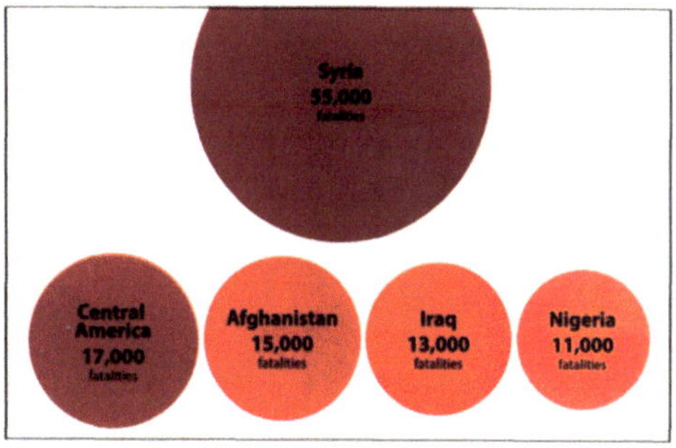

Shia militias in Iraq and Syria

There are more than 40 militias in Iraq that form the Hashd al-Shaabi (Popular Mobilisation Units, or PMU), and that fight against the Islamic State (also known as ISIS or ISIL). Most of these militias are Shia, but some Sunnis and members of other sects also participate.

The PMU is hardly cohesive. There have been cases of infighting between the militias, clashes between the militias and Iraqi security forces, and battles between the militias and the Kurdish Peshmerga. Although they are technically under the Iraqi government's command, the groups are ideologicall diverse: some are linked to Iraqi political parties or are closely aligned with Iran, while others formed after Grand Ayatollah A al-Sistani issued in June 2014 a fatwa that called for Shi'ites to fight against ISIS.

Many have increasingly fought alongside the forces of President Bashar al-Assad in Syria, along with Iranian forces and other Shia militias from elsewhere in the region. In Syria, they have primarily fought against rebels, rather than ISIS.

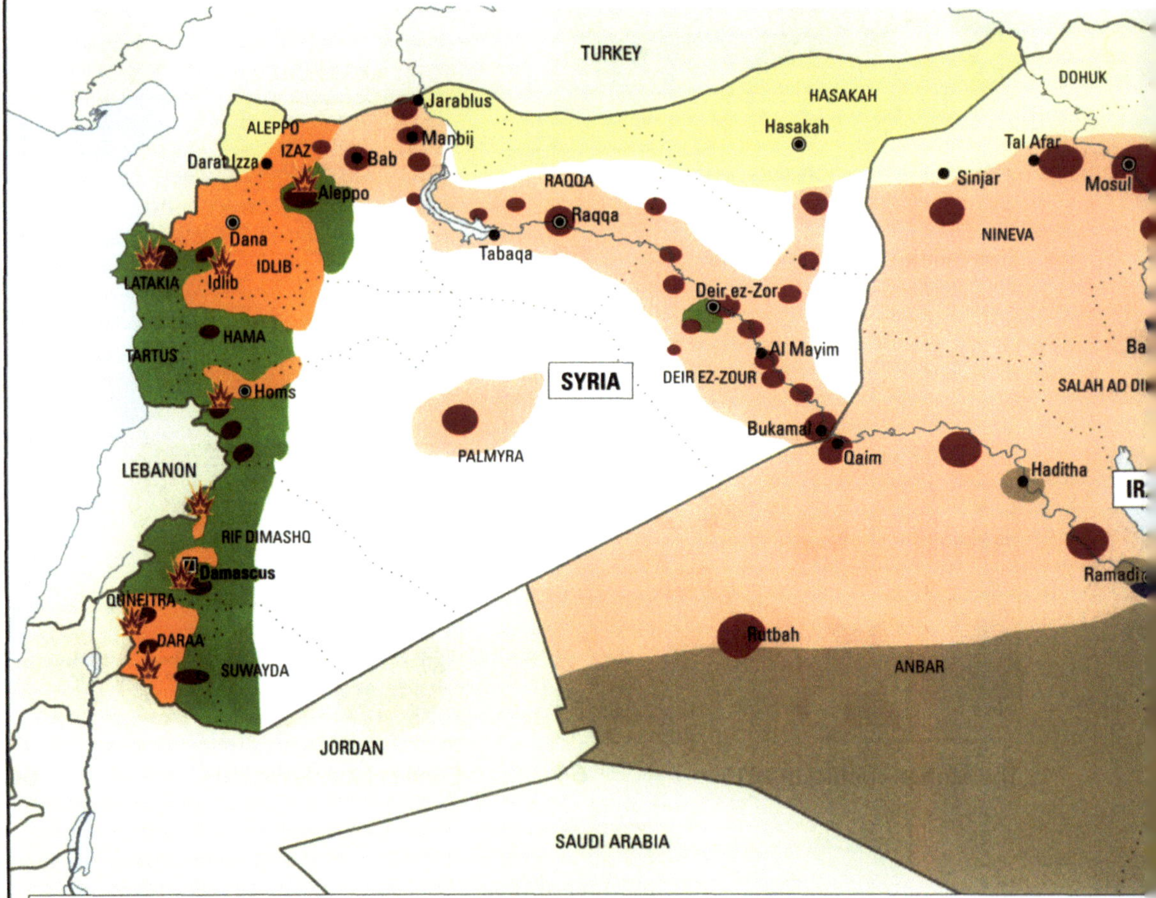

Shia militias in Syria

IRGC Quds Force
Country: Iran
Leader: Qassem Suleimani
Orientation: special forces as well as regular IRGC units
Size: 2,000 present in Syria

Hizbullah
Country: Lebanon
Leader: Hassan Nasrallah
Orientation: officially announced its involvement in Syria in 2013 in the battle for Qusair; funded by and close to Iran
Size: 5,000–8,000 in Syria

Liwaa Fatimiyoun
Country: Afghanistan
Leader: unknown (was Alireza Tavassoli; killed in 2015)
Orientation: formed in 2014; established by and report to IRGC
Size: a few hundred

Liwaa Zaynabiyoun
Country: Pakistan
Leader: unknown
Orientation: formed in 2014; established by and reports to the IRGC; some involvement in Iraq
Size: unknown

Liwaa Abul Fadl al-Abbas
Country: Syria/Iraq
Leader: Abu Ajeeb and Sheikh Izzedine al-Darraji
Orientation: established in 2012 proclaiming to protect Shia holy sites in Syria; several other smaller branches have emerged from Liwaa Abul Fadl al-Abbas; also has Iraqi branches: Qaeda Quwet Abu Fadl al-Abbas operating in Iraq under Aud al-Khafaji and Abu Kamil al-Lami, and Liwa Abu Fadhl al-Abbas–Tashkil al-Iraq under Qassem al-Tai; primarily Iraqi fighters; also present in Iraq since 2014 (such as in Samarra and Anbar) under the name Liwaa Abul Fadhl al-Abbas Tashkil al-Iraq.
Size: a few thousand

Liwaa Zhulfaqqar
Country: Syria/Iraq
Leader: Abu Shahed al-Juburi
Orientation: formed in 2013 as a part of the Liwaa Abul Fadl al-Abbas network, also involved in Iraq (s as in Diwaniya)
Size: 1,000

Liwaa al-Imam Hussein
Country: Iraq/Syria
Leader: Sheikh Ab Karrar al-Bahladi
Orientation: another group wit links to Liwaa Abul Fadl al-Abb Sadrist, links with Iran
Size: unknown

Liwaa al-Imam Has al-Mujtaba
Country: Iraq/Syria
Leader: Sheikh Ahram Kaabi (c

Sources: IISS research, Institute for the Study of War (ISW), Thomas van Linge, Phillip Smyth (Washington Institute for Near East Policy)

Shia militias in Iraq

Prime Minister Haidar al-Abadi

Iraqi security forces

Popular Mobilisation Units
[around 120,000 fighters] coordinated by Faleh al-Fayad

Kata'ib Hizbullah
Leader: Jamal al-Ibrahimi (aka Abu Mahdi al-Muhandis)
Orientation: founded in 2007, ideologically and politically aligned with Iran; also has smaller groups operating under it, such as Saraya al-Difa' al-Shaabi, which has been present in Syria since at least 2013
Size: 2,000–10,000
Involvement in Syria: present throughout Syria

Asa'ib Ahl al-Haq
Leader: Qa'is al-Khaz'ali
Orientation: founded in 2006, splintered from Jaysh al-Mahdi after it was frozen by Sadr; close to Iran
Size: 1,000–5,000
Involvement in Syria: present through front groups such as Liwaa Kafeel Zeinab and Haidar al-Karar

Badr Organisation
Leader: Hadi al-Ameri
Orientation: established in Iran in 1980s, returned to Iraq after 2003, close to Iran
Size: Up to 20,000
Involvement in Syria: Damascus, Aleppo

Saraya al-Salam
Leader: Muqtada Sadr
Orientation: formed in 2014 as a successor to Sadr's Jaysh al-Mahdi and Liwaa al-Yawm al-Maw'ud. Allied with Iraqi government but still linked to Iran
Size: 15,000 (2008)
Involvement in Syria: none

Kata'ib al-Imam Ali
Leader: Shibl al-Zaydi
Orientation: formed in 2014; armed wing of Harakat al-Iraq al-Islamiya
Size: unknown
Involvement in Syria: Damascus

Kata'ib Sayyid al-Shuhadaa
Leader: Abu Mustafa al-Sheibani
Orientation: founded in 2013, works with Liwaa Abul Fadl al-Abaas; links with Badr, close to Iran
Size: less than 500
Involvement in Syria: southern Damascus, eastern Ghouta

Harakat Hizbullah al-Nujaba
Leader: Sheikh Ahram Kaabi
Orientation: formed from Asa'ib Ahl al-Haq in 2013 to participate in Syria (but redeployed to Iraq in 2014); a front group for Nujaba Movement is Liwaa Ammar Ibn Yassir, Liwaa al-Imam Hassan al-Mujtabi, Liwaa al-Hamad; closely aligned with Iran
Size: a few thousand
Involvement in Syria: Aleppo, Hama, Damascus, Latakia, Idlib

Saraya Tala'e' al-Khorasani
Leader: Ali al-Yaseri
Orientation: formed in 2013 to fight in Syria, direct ties with Iran, deployed to Iraq after fall of Mosul in 2014
Size: 3,000
Involvement in Syria: eastern Ghouta, Rif Dimashq, Damascus airport

Kata'ib Jund al-Imam
Leader: Abu Jaafar al-Asadi
Orientation: founded in 2014, close to Iran
Size: unknown
Involvement in Syria: unclear

Saraya Ashoura
Leader: Ammar al-Hakim
Orientation: founded in 2014, close to Iraqi party Supreme Islamic Council, pro-Iran
Size: unknown
Involvement in Syria: unclear

Saraya al-Jihad
Leader: Hassan al-Sari
Orientation: formed in 2014, linked to Iraqi party Harakat al-Jihad wal-Binaa, pro-Iran
Size: around 3,000
Involvement in Syria: unclear

Kata'ib al-Taiar al-Resali
Leader: Sheikh Adnan al-Shahmani
Orientation: formed in 2014 after Sistani's fatwa; affiliated with Iraqi party al-Tayar al-Risali
Size: unknown
Involvement in Syria: unclear

Saraya Ansar al-Aqeeda
Leader: Sheikh Jalal al-Din al-Saghir
Orientation: formed in 2014; has links to Badr and ISCI
Size: unknown
Involvement in Syria: unclear

Faylaq Waad al-Sadiq
Leader: Sheikh Mohammed Hamza al-Tamimi
Orientation: thought to have been established in 2010 but was relatively inactive until 2012–13
Size: 1,000
Involvement in Syria: Aleppo

Harakat Ansar Allah al-Awfiyya
Leader: Haidar al-Gharawi
Orientation: formed in 2014 from Kayan al-Sadiq wal 'Atta fi Iraq party from Maysan province, split from Harakat al-Nujaba; Khomeinist
Size: unknown
Involvement in Syria: close links with Liwaa al-Hamad, another Harakat al-Nujaba group

…al-Nujaba) and Thamer …di
…tion: one of Harakat …ba's fronts for participation …; has many subunits, …g Kata'ib al-Ashtar, Kata'ib …mr and Saraya al-Shahid …Kayara
…00 (including Liwaa Ammar …sir, Liwaa al-Hamad)

Liwaa Ammar Ibn Yassir
Country: Iraq/Syria
… Sheikh Ahram Kaabi (of … al-Nujaba)
…tion: founded in 2013; one of … al-Nujaba's fronts for participation in Syria
Size: 1,500 (including Liwaa al-Imam Hassan al-Mujtaba and Liwaa al-Hamad)

Liwaa al-Hamad
Country: Iraq/Syria
Leader: Sheikh Ahram Kaabi (of Harakat al-Nujaba)
Orientation: founded in 2013; one of Harakat al-Nujaba's fronts for participation in Syria
Size: 1,500 (including Liwaa al-Imam Hassan al-Mujtaba and Liwaa Ammar Ibn Yassir)

Liwaa Assad Allah al-Ghalib
Country: Iraq/Syria
Leader: Abu Fatima al-Moussawi
Orientation: pro-Iran; Sadrist; also has a group operating in Iraq: Kata'ib Assad Allah al-Ghalib, led by Abu Ghayth al-Hassani
Size: a few hundred

Kata'ib Sayyid al-Shuhadaa
Country: Iraq/Syria
Leader: Abu Mustafa al-Sheibani
Orientation: founded in 2013; links with Badr; close to Iran; also involved in Iraq
Size: a few hundred

Liwaa Kafeel Zeinab
Country: Iraq/Syria
Leader: unknown; possibly Muhammad al-Tabatabai
Orientation: Asa'ib Ahl al-Haq's front for participation in Syria; another group is Haidar al-Karar (an 800-man force that operates in Aleppo)
Size: unknown; mostly Asa'ib Ahl al-Haq personnel

Quwwat al-Ridha
Country: Syria
Leader: unknown
Orientation: one of many Hizbullah-affiliated groups that have sprung up in Syria; another is al-Muaqma al-Wataniyya al-Aqadiyya. Quwwat al-Ridha is also affiliated with the National Defence Force
Size: unknown

Afwaj al-Kafeel
Country: Iraq/Syria
Leader: Ahmed Haji al-Saadi; redeployed to Iraq to participate in Fallujah operations
Orientation: linked to Liwa Abu Fadl al-Abbas
Size: unknown

Decline in conflict between Colombian government and guerrillas during FARC peace talks

Number of offensive actions by armed groups in Colombia against government and civilian targets

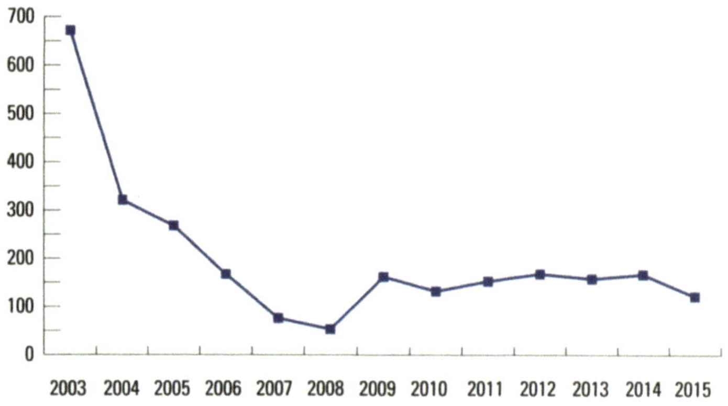

Offensive actions by armed groups during peace talks with FARC 2014–15

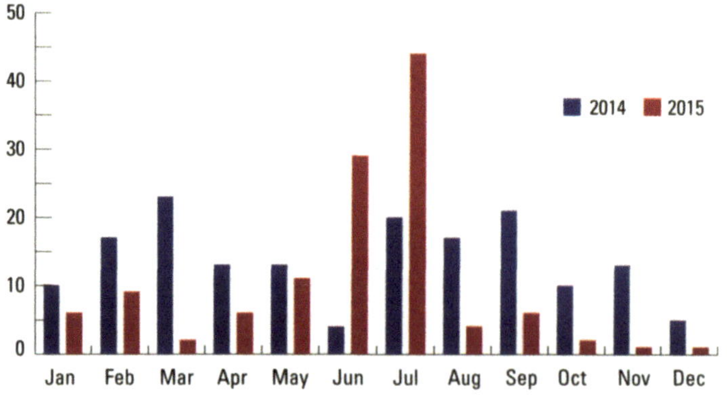

May 2014 — **January 2015**

16 May 2014: Negotiators in Havana announce that a deal has been reached on the third item on the agenda, regarding drug trafficking and the cultivation of illegal plants.

17 November 2014: President Juan Manuel Santos announces the suspension of the peace talks after FARC captures a Colombian general in Choco Department.

30 November 2014: FARC frees General Ruben Dario Alzate, whose capture had caused the government to interrupt peace talks.

20 December 2014: FARC's unilateral ceasefire starts.

7 March 2015: Negotiators announce a cooperation agreement for the demining of Colombian territory.

15 April 2015: Eleven Colombian soldiers are killed during a FARC ambush in Cauca. President Santos ends the moratorium on airstrikes.

May 2015

22 May 2015: A government airstrike kills 26 FARC fighters in Cauca Department. In response, FARC suspends its unilateral ceasefire.

25 June 2015: A FARC attack on 8 June against a section of the Transandino oil pipeline causes the country's 'worst environmental damage in ten years', says the government.

20 July 2015: FARC starts a new unilateral ceasefire.

23 September 2015: Government and FARC reach a preliminary agreement on transitional justice and the establishment of a special tribunal.

01 October 2015: FARC' leader Rodrigo 'Timochenko' Londoño orders all of the group's members to cease military training.

15 December 2015: Negotiators announce a deal on the fourth point of the agenda: reparations for victims and transitional justice.

Sources: IISS Armed Conflict Database, Instituto de Estudios para el Desarollo y la Paz (Indepaz), Ministry of Defence of Colombia, National Police of Colombia

The two most powerful criminal groups in Colombia are Los Urabeños and Los Rastrojos, both of which developed out of right-wing paramilitary groups that demobilised during the 2000s. Many of the country's gangs were formed or grew quickly after the demobilisation, and some established themselves as nationwide enterprises. The groups are seen as potential threats to stability in the areas of Colombia currently occupied by FARC.

Number of municipalities in which criminal groups are present

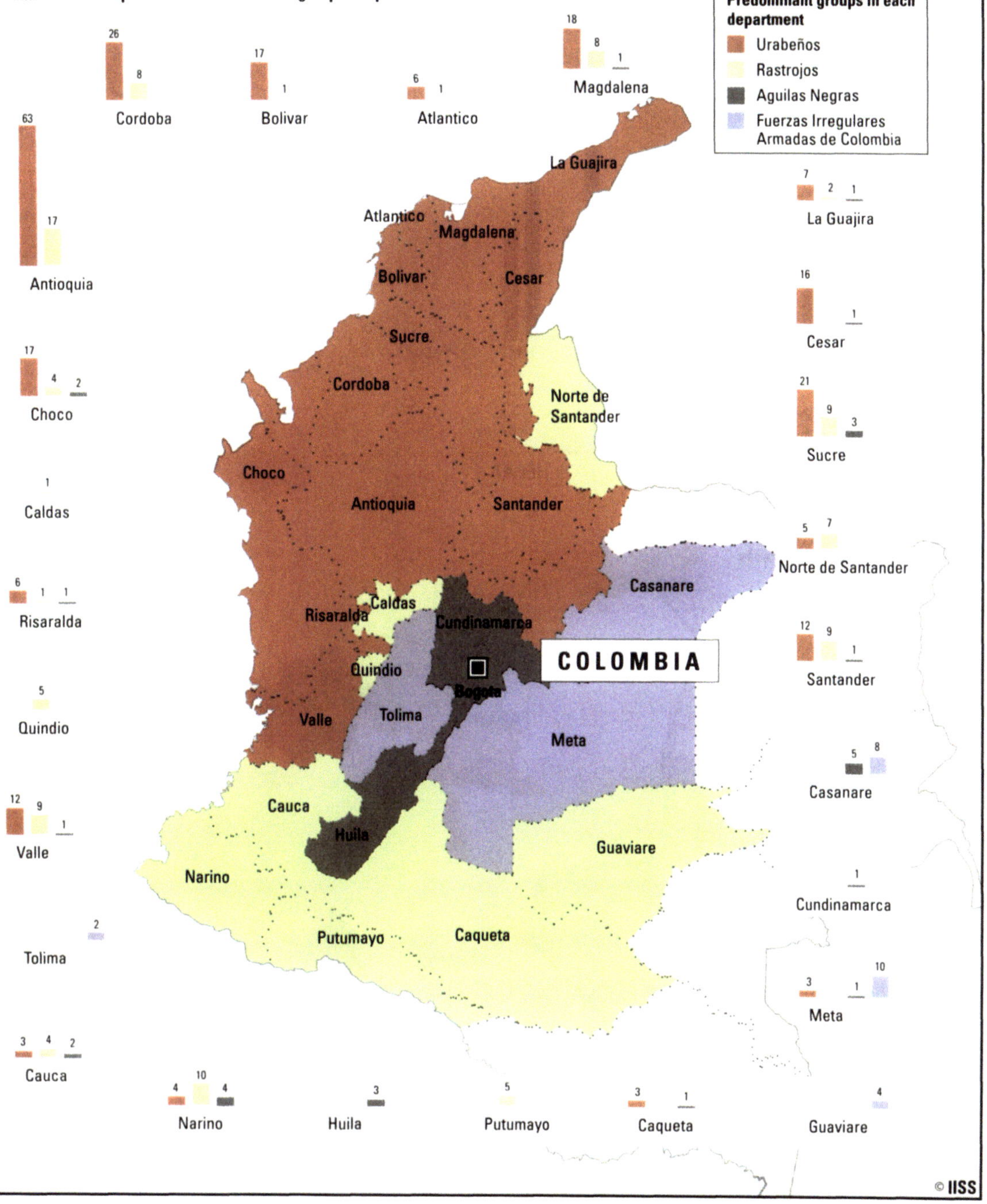

The Afghan conflict in 2015

The conflict in Afghanistan worsened in 2015. The Taliban was emboldened by the full transfer of security responsibilities to Afghan forces, and NATO's International Security Assistance Force (ISAF) was folded into the *Resolute Support* train, advise and assist mission, with more limited rules of engagement. *Operation Freedom's Sentinel* was launched as a separate US counter-terrorism mission. At the same time, military operations in Pakistan's tribal areas pushed more militants into Afghanistan, and an offshoot of the Islamic State (also known as ISIS or ISIL) took territory in eastern Afghanistan. Conflict fatalities increased sharply from 2014, as the UN assessed the Taliban's presence to be more widespread than at any time since 2001. The peace process was disrupted by Taliban disunity after the death of Mullah Omar. Recognising Afghan forces' continued need for strong support, the United States and other NATO allies decided to slow their withdrawal of troops.

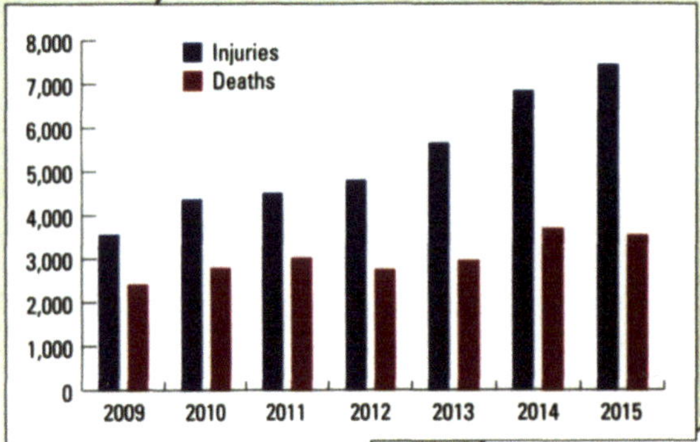

Civilian casualties
Figures collected by the United Nations Assistance Mission in Afghanistan (UNAMA) showed a steady rise in civilian casualties; the vast majority (62% in 2015, down 10% from 2014) were caused by insurgents, particularly their use of improvised explosive devices and suicide bombers.

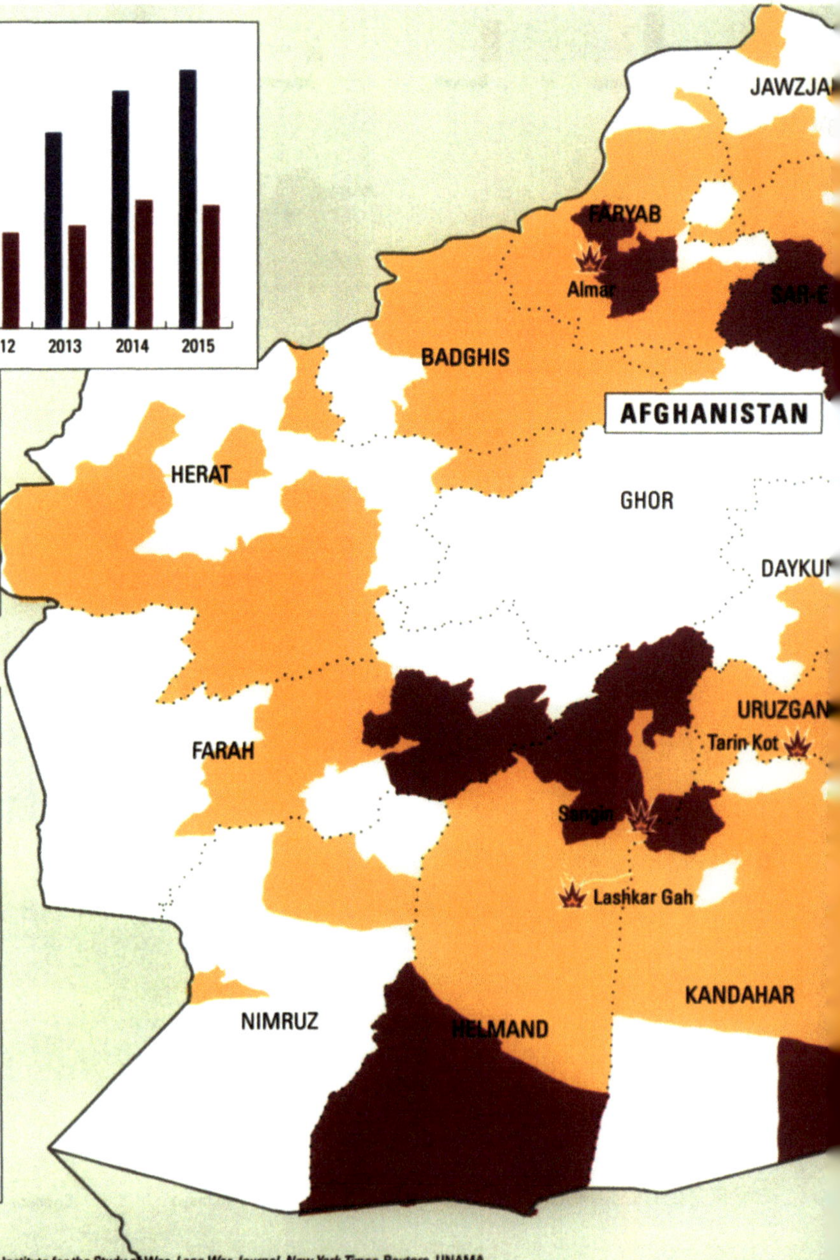

The death of Mullah Omar
The 29 July revelation that Taliban leader Mullah Omar had died in 2013 had a significant effect on Taliban unity, and on the nascent peace process. A second round of unprecedented official talks between the Afghan government and the Taliban was cancelled, and dialogue did not resume in 2015. The Taliban nearly fractured after disagreements over the selection of Mullah Mohammad Akhtar Mansour as its new leader, and clashes between rival factions occurred regularly over the second half of the year. By the end of 2015, Mansour appeared to be consolidating his hold on the organisation. Aided by growing Chinese engagement, efforts were under way to create conditions for talks in 2016.

Sources: AFP, BBC, IISS Armed Conflict Database, Institute for the Study of War, *Long War Journal*, *New York Times*, Reuters, UNAMA

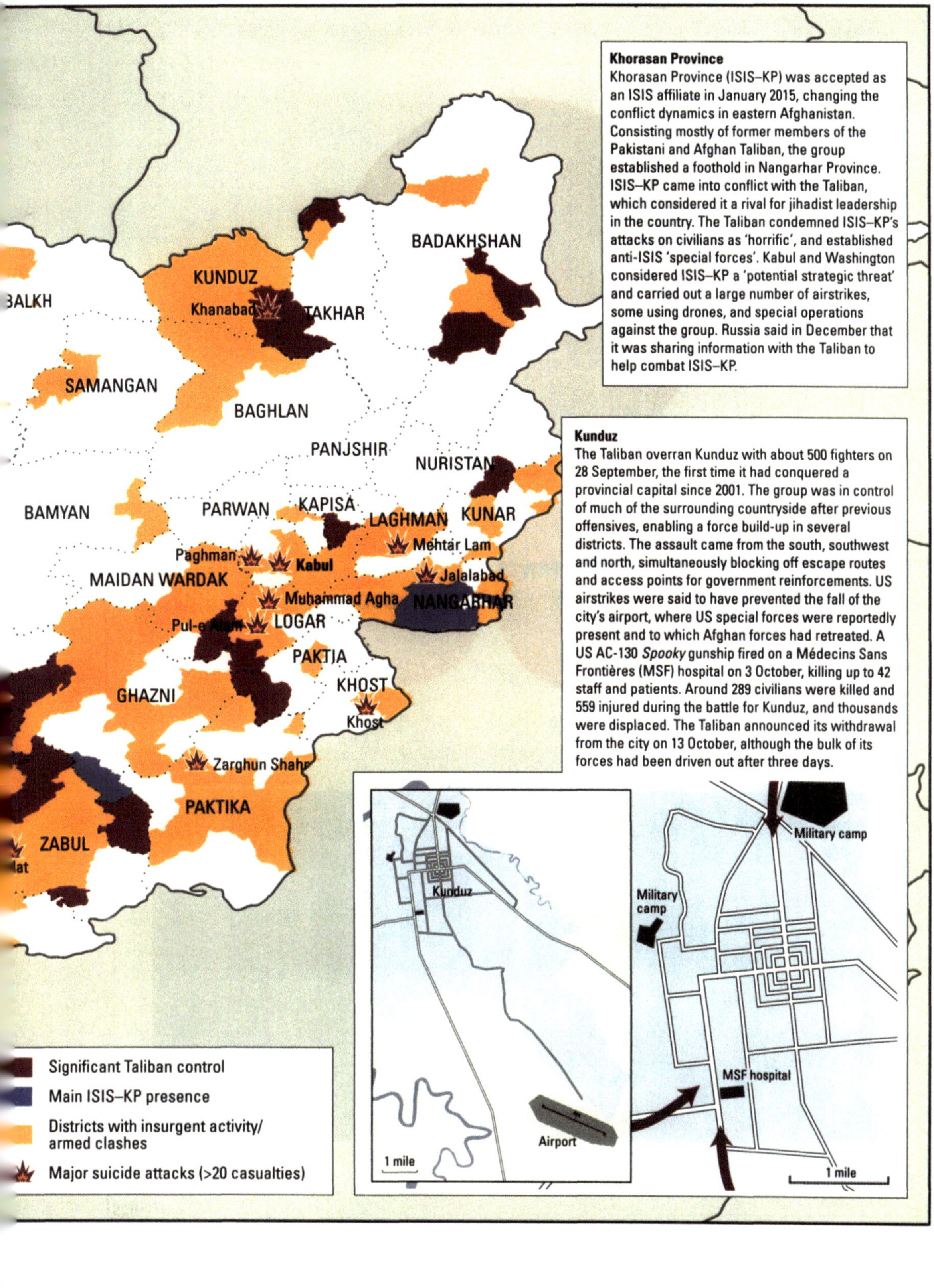

Khorasan Province

Khorasan Province (ISIS–KP) was accepted as an ISIS affiliate in January 2015, changing the conflict dynamics in eastern Afghanistan. Consisting mostly of former members of the Pakistani and Afghan Taliban, the group established a foothold in Nangarhar Province. ISIS–KP came into conflict with the Taliban, which considered it a rival for jihadist leadership in the country. The Taliban condemned ISIS–KP's attacks on civilians as 'horrific', and established anti-ISIS 'special forces'. Kabul and Washington considered ISIS–KP a 'potential strategic threat' and carried out a large number of airstrikes, some using drones, and special operations against the group. Russia said in December that it was sharing information with the Taliban to help combat ISIS–KP.

Kunduz

The Taliban overran Kunduz with about 500 fighters on 28 September, the first time it had conquered a provincial capital since 2001. The group was in control of much of the surrounding countryside after previous offensives, enabling a force build-up in several districts. The assault came from the south, southwest and north, simultaneously blocking off escape routes and access points for government reinforcements. US airstrikes were said to have prevented the fall of the city's airport, where US special forces were reportedly present and to which Afghan forces had retreated. A US AC-130 *Spooky* gunship fired on a Médecins Sans Frontières (MSF) hospital on 3 October, killing up to 42 staff and patients. Around 289 civilians were killed and 559 injured during the battle for Kunduz, and thousands were displaced. The Taliban announced its withdrawal from the city on 13 October, although the bulk of its forces had been driven out after three days.

Conflict fatalities in 2015

The Middle East accounted for just under half of conflict fatalities worldwide in 2015, according to Armed Conflict Database figures.

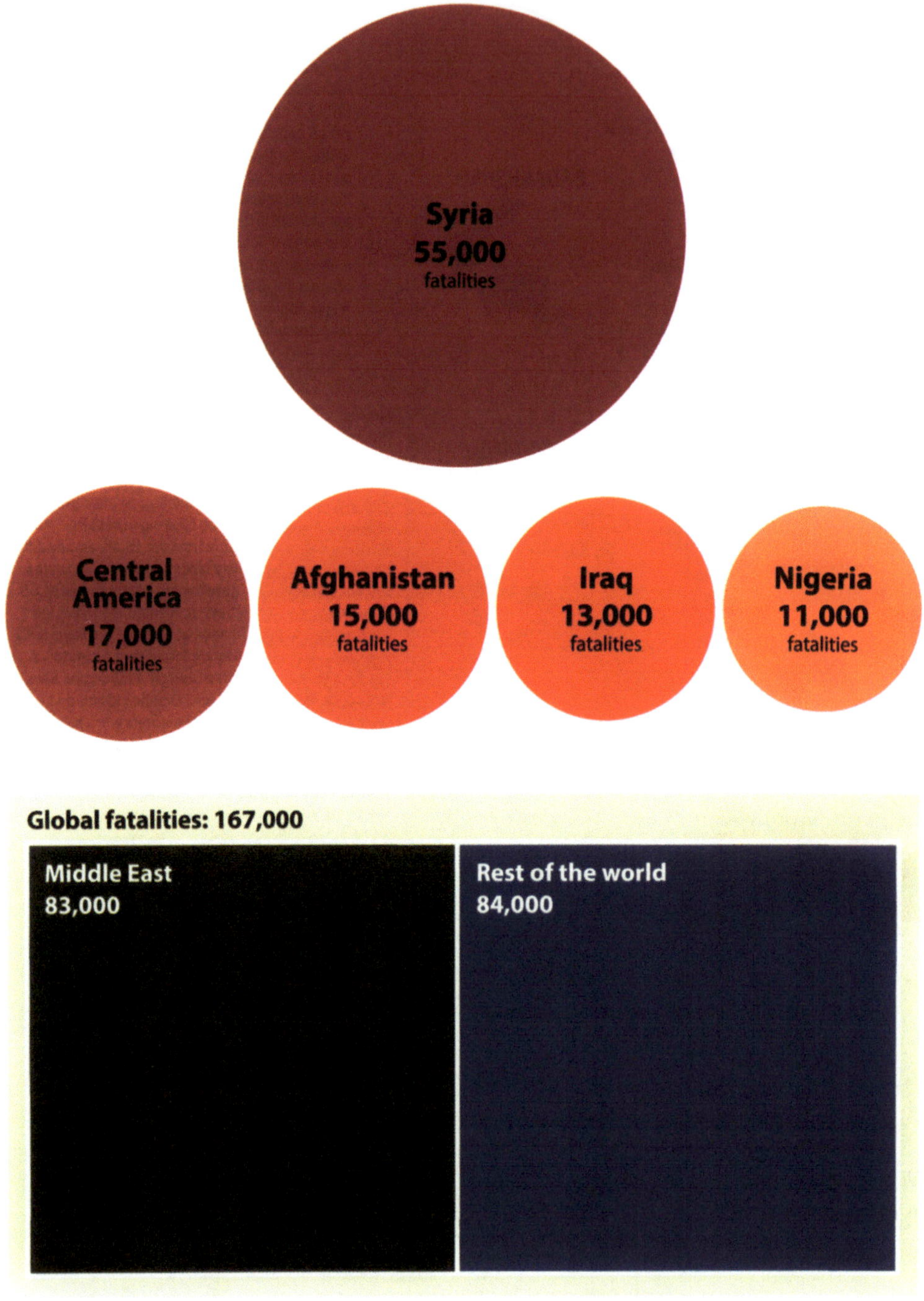

Chapter Three

Middle East

Egypt (Sinai)

Key statistics	2014	2015
Conflict intensity:	Medium	Medium
Fatalities:	900	3,000
New IDPs:		
New refugees:		

As President Abdel Fattah al-Sisi strengthened his control of Egypt throughout 2015, the security situation in the country deteriorated significantly. Wilayat Sina (Sinai Province), an insurgent group that pledged allegiance to the Islamic State (also known as ISIS or ISIL) in November 2014, posed a severe security threat, launching deadly attacks and challenging the security forces' tenuous control over Sinai. Egypt concurrently faced a medium-level insurgency led by radicalised individuals linked to the Muslim Brotherhood. The complex security environment was used by the Sisi-led government as a pretext for restricting space for political participation. Many of the government's policies, however, appeared to be slowly generating discontent and undermining the president's support base.

Escalation in Sinai

Formerly known as Ansar Bayt al-Maqdis (ABM), Sinai Province evolved into the main non-state armed group active in the Sinai Peninsula after it subsumed

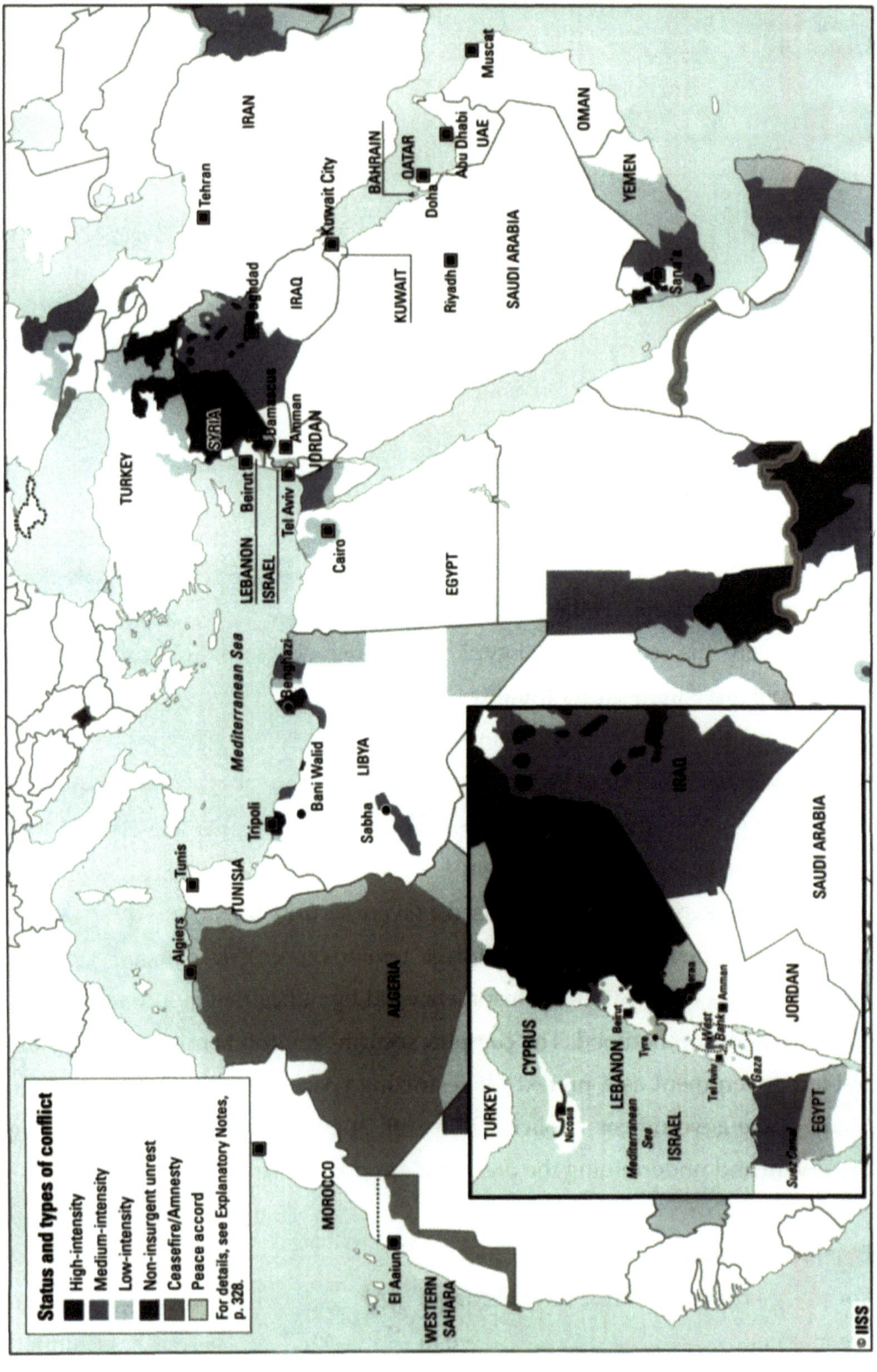
Status and types of conflict
High-intensity
Medium-intensity
Low-intensity
Non-insurgent unrest
Ceasefire/Amnesty
Peace accord
For details, see Explanatory Notes, p. 328.
WESTERN SAHARA
El Aaiun
MOROCCO
ALGERIA
Algiers
Tunis
TUNISIA
Tripoli
Bani Walid
Sabha
LIBYA
Benghazi
Mediterranean Sea
LEBANON
ISRAEL
Beirut
Tel Aviv
Cairo
EGYPT
TURKEY
SYRIA
Damascus
Amman
JORDAN
Baghdad
IRAQ
KUWAIT
Kuwait City
Riyadh
SAUDI ARABIA
Tehran
IRAN
BAHRAIN
QATAR
Doha
Abu Dhabi
UAE
Muscat
OMAN
YEMEN
Sana'a
TURKEY
CYPRUS
Nicosia
LEBANON
Beirut
Tyre
Mediterranean Sea
ISRAEL
Tel Aviv
West Bank
Amman
Gaza
Suez Canal
EGYPT
JORDAN
IRAQ
SAUDI ARABIA
© IISS

smaller *takfiri* groups (a *takfiri*, in this context, is an extremist who accuses other Muslims of apostasy and engages in armed activity). Since Sisi's rise to power, in July 2013, Sinai Province has targeted the security forces with renewed vigour in northeastern and central Sinai, and in 2015 it expanded its capacity to launch increasingly complex attacks, including those outside its traditional area of operation.

The group's largest attack, carried out on 31 October, was in many ways dissimilar to its past operations: one of its operatives placed an explosive device on board a Russian commercial plane, Metrojet 9268, killing all 217 passengers and seven crew members shortly after the jet departed from Sharm el-Sheikh airport. The incident amounted to a qualitative escalation in the conflict, since Sinai Province had previously focused on targeting the security forces and eschewed publicising attacks on civilians. In the wake of the attack, the hitherto obscure group gained international attention that focused on the extent of its coordination with ISIS. The incident confirmed long-standing speculation that Sinai Province has progressively embraced aspects of the modus operandi of ISIS since pledging its allegiance to the group – proving, in the words of analyst Mokhtar Awad, that 'the alliance was more than just rhetorical'.

The operation was largely perceived as a form of retaliation for Russian airstrikes against ISIS in Syria – and ISIS presented it as such – but it also corresponded with Sinai Province's declared strategy of 'economic warfare' against the state. Infiltrating Sharm el-Sheikh, a major tourist hub on the southern tip of the peninsula, was highly symbolic: the city was the venue for a large investment conference convened the previous March, which reflected its potential role as a major source of income.

In the cities of Arish, Rafah and Sheikh Zuweid, Sinai Province increasingly launched large-scale attacks on multiple fronts, relying on a combination of capabilities, including vehicle-borne explosive devices, suicide bombs, mortars and small arms. Such high-visibility attacks serve the insurgents' goal of undermining the military's presence and discrediting its claim to being in control. The release of vast amounts of audio and visual material – showing footage of attacks that contradicted the army's version of events, and displaying new weaponry — was another aspect of this strategy.

These events formed part of a consistent trend of escalation by both sides in the conflict. After an attack in late January killed 32 soldiers, Sisi established a unified military command east of the Suez Canal to oversee the counter-terrorism activities of the Second and Third armies. But the move did not curtail militant activity, and Sinai Province subsequently claimed responsibility for several other large attacks. Its operatives stormed Sheikh Zuweid in early July and prevented the military from entering the city. They only retreated ten hours later, under heavy aerial fire. The attack functioned as a show of force, and led to widespread panic across the entire country. The army subsequently launched *Operation Martyr's Right*, an intensive campaign that reportedly killed 500 suspected militants in the first 16 days alone. The military deemed this first phase a resounding success, and announced the beginning of the 'second phase' focusing on stabilisation, but militant activity continued unabated.

Despite a constant surge in troop deployment, the military largely maintained set positions in North Sinai, and the few movements it made were predictable, especially compared to the fluctuating and dynamic movements of the insurgents. At the same time, the army has only a weak capacity for counter-insurgency operations, due to its lack of intelligence networks. The media reported in May that the military had enlisted North Sinai's largest tribes in the fight against *takfiri* groups, but actual clashes between the militants and the tribes were mostly exaggerated. In fact, as most of the prominent tribal leaders who voiced support for military operations were based in Cairo, the reports can be dismissed as another product of the propaganda war between the military and the insurgents.

Sinai Province aimed to balance its alliance with ISIS (which may be motivated by a desire for the latter's resources) with its relationship with residents of North Sinai, among whom it has sought to cultivate legitimacy as an alternative to the state. Sinai Province's effort to gain local support can be discerned from its statements emphasising the government's unjust policies on North Sinai's long-marginalised population, and the military's violence against women in the area and beyond. Yet despite such declarations, the group's ISIS-style attacks estranged potential support bases. It has killed more than 130 residents accused of collaborating with the security forces. Sinai Province even claimed responsibility for assassinating a Cairo-based leader of the Sawarka tribe who supported the army, demonstrating both the extent of its threat to alleged collaborators and its reach beyond the peninsula.

Thus, residents of North Sinai found themselves stuck between the threat from armed groups, primarily Sinai Province, and that from the military's 'scorched earth' policies. Residents of the area accused the military of subjecting them to collective punishment, compounding decades of mutual mistrust and the government's perception of the local Bedouin population as disloyal. Civilians were killed, injured and extrajudicially detained, while the homes of some were demolished without official acknowledgement or adequate compensation. The army also proceeded with a controversial plan to level thousands of homes in Rafah, clearing the way for the construction of a buffer zone along the border with the Gaza Strip.

Nationwide insurgency

Both Sinai Province and ISIS itself claimed responsibility for a number of attacks against strategically important infrastructure in Cairo, including a security building and a courthouse in Shubra al-Khaymah, a building affiliated with the Egyptian foreign ministry, and the Italian consulate compound. The bombings resulted in few casualties, but the militants subsequently declared that such 'security buildings' were legitimate targets. The ability of ISIS to infiltrate Cairo and attack state and military positions is contingent upon its cooperation with smaller groups on the ground – possibly an undeclared proxy independent of its Sinai affiliate.

In addition to cooperating with ISIS, Sinai Province joined forces with cells outside of the peninsula and cooperated with armed groups to extend its reach throughout the country. Sinai Province took responsibility for the abduction and subsequent beheading of Croatian hostage Tomislav Salopek in August, but it was a Cairo-based cell that had kidnapped him on a highway west of the city. Sinai Province initially threatened to behead Salopek unless Sisi released 'all jailed Muslim women' within 48 hours. The announcement was partly an attempt to dispel the hype surrounding the opening ceremony for the 'new Suez Canal', scheduled to take place the following day, and to thereby embarrass Sisi. A key figure involved in Salopek's abduction was Ashraf al-Gharabli, who was killed by the Egyptian security forces in Cairo in November. He had reportedly coordinated militant activity in the Nile Valley and mediated Sinai Province's entry into mainland Egypt.

Another link between Sinai Province and mainland Egypt is Ajnad Misr, which formed when its leader, Hammam Mohammed (also known as Majd al-Din al-Masri), left ABM and Sinai to establish an armed wing closer to the capital. Ajnad Misr claimed responsibility for relatively small-scale attacks in the Greater Cairo area, which targeted security personnel in retaliation for what it called the government's 'illegitimacy' and 'tyranny'. The security forces killed Mohammed in Giza in April. This prompted eulogies from al-Qaeda franchises such as al-Qaeda in the Arabian Peninsula and al-Qaeda in the Islamic Maghreb, which confirmed speculation that he remained loyal to al-Qaeda after ABM pledged allegiance to ISIS. Although Sinai Province's pledge of allegiance to ISIS reportedly alienated figures within the group and led to splits, the rivalry between ISIS and al-Qaeda was not as strong in Egypt as elsewhere in the region. Rather, these divisions seemed remarkably fluid on the ground, with reports of cross-organisational collaboration.

Furthermore, there were indications that former military personnel had participated in the insurgent violence in Sinai and elsewhere in Egypt. These officers have insider knowledge of the military's training and combat methods, and are thus particularly effective at prosecuting the insurgency. Hisham al-Ashmawy, a former special-forces officer hiding in Libya, announced in July the establishment of insurgent group al-Mourabitoun and urged his followers to attack Sisi, whom he described as a new pharaoh. Ashmawy had previously served as the head of Sinai Province's 'military training committee', but split from the group after it switched allegiance from al-Qaeda to ISIS. Egyptian security sources have implicated him in several assassinations, and believe that he masterminded the most high-profile such attack in the history of the insurgency: the killing of chief prosecutor Hisham Barakat in June. But responsibility for the assassination was claimed by the Popular Resistance Movement, which is affiliated with the Muslim Brotherhood and primarily operates in Giza. It is unclear whether the group has received orders or guidance from Ashmawy.

Continued repression of the Muslim Brotherhood in Egypt

Barakat's assassination resulted from long-developing resentment of the judiciary, particularly its facilitation of the government's efforts to stamp out the Brotherhood. He gained notoriety by handing down mass death sentences to hundreds of Brotherhood members and supporters, including former president Muhammad Morsi. An array of other Brotherhood leaders, among them the general guide and his deputy, received similarly harsh sentences in trials that were considered by most observers to be politically motivated. The Brotherhood's members and supporters continued to be arrested, with the police violently dispersing their protests.

Emerging amid these restrictions, the Popular Resistance Movement has contributed to the rising insurgency. After the Brotherhood was declared an illegal terrorist organisation, in 2014, some of its members, particularly young people, began to increasingly resort to violence. They split from the National Alliance to Support Legitimacy, an Islamist umbrella group that opposed the military *coup d'état* of 2013, and staged frequent attacks against police and public facilities using improvised explosive devices and small arms. Their targets have included banks, electricity infrastructure and occasionally even civilian spaces such as restaurants. The group does not have an organised leadership, but its ad hoc structure appeals to and encourages lone-wolf attacks.

The Popular Resistance Movement is on one side of a fissure within the Brotherhood between youth cadres advocating a revolutionary direction, which entails a bolder confrontation with the government, and the senior leadership, which still favours a non-violent approach. The Brotherhood's leaders, especially those in exile, have called for peaceful protests and civil-disobedience campaigns, and have focused on internationalising their struggle. Nevertheless, they have issued statements that implicitly encourage violent acts. After the security forces killed 13 Brotherhood figures in a raid in July, the leadership called for a revolt. Morsi similarly called on his supporters to save the country from the 'oppressive junta', signing his message as the 'president of the Arab Republic of Egypt'.

The spread of militant activity from Sinai to the rest of the country combined with rising violence by disaffected members of the Brotherhood, particularly in large urban areas, to drive the insurgency. The progressive entanglement of these two strands of violence lent credence to the government's claim that they were inherently connected. This potentially functioned as a self-fulfilling prophecy, since the government had linked the threats since the start of the 'war on terror', in 2013.

Restrictive political climate

The increasingly restrictive political climate in Egypt can be seen as both a cause and a consequence of the pronounced violence. In the absence of a parliament, policies were mostly implemented via presidential decree. Although parliamentary elections were finally held in November–December 2015, following a two-year delay, they were largely inconsequential, with pro-Sisi blocs winning the majority of seats. A turnout of slightly more than 25% suggested that most Egyptians were indifferent to the process, and were aware that parliament would play only a symbolic role alongside the overwhelmingly powerful executive branch. Sisi and his allies suggested that the president will be granted even more power, at the expense of parliament. Sisi has suggested that the parliament should give up some its constitutional powers, which include restricting the president's authority in choosing the prime minister and in undertaking loans. Such measures were justified as steps towards unity, in the face of militant groups' efforts to destabilise the state.

Counter-terrorism legislation was used to further undermine political participation, building on previous measures such as the 2013 protest law and a 2014 decree that placed all critical public facilities under the army's jurisdiction. A law enacted in August 2015 granted the security forces protection from prosecution, and gave prosecutors new surveillance and detention powers. It also facilitated expedited trials in special courts for terrorism-related cases, and banned media reports that contradicted the official account of terrorist attacks. An earlier draft of the bill specified that those convicted of writing such reports would receive jail sentences of up to two years, before this was amended to a fine of US$25,000–65,000.

The law aims to strengthen the credibility of the government and the armed forces against any reports that may undermine it. The foreign ministry lashed out at the press for suggesting that it was plausible Metrojet 9268 had been brought down by a bomb, a theory that the government later officially rejected. Egypt was at that point still recoiling from another, embarrassing incident, in which the air force killed 12 tourists, including eight Mexicans, in the Western Desert. The new legislation illustrates not only the government's hostility towards journalists and dissidents, but also its willingness to link dissent with terrorism. The number of journalists in Egyptian jails reached an all-time high in 2015, despite the widely publicised release of three Al-Jazeera English journalists, under immense international pressure.

While there was growing evidence of domestic discontent with Sisi, the international community seemed to swing in the opposite direction. There were positive overtures to his regime from Western countries, likely due to the array of security challenges facing Egypt. The US incrementally reversed a ban on military aid to the country in place since 2013, which had led the Cairo to search for alternative suppliers. Washington's monopoly on providing weapons to Egypt officially ended in February 2015, when Egypt received 24 fighter jets from France. Cairo signed another large contract with Paris later in the year, and more deals with various European countries, as well as Russia, were revealed.

However, the majority of Egypt's new weapons are designed more for traditional warfare rather than counter-insurgency, suggesting that Cairo has ambitions for a more active military role in the wider Middle East. Critics described this intensive militarisation programme as a waste of resources, pointing to both Egypt's deteriorating socio-economic conditions and the military's involvement in the economy, which unofficial estimates place at around 40% of GDP. The programme may be another attempt at state self-aggrandisement, and a continuation of ambitious and nationalistic projects such as those to expand the Suez Canal and to construct a 'new Cairo' east of the original city.

The insurgencies raging throughout Egypt led to a sharp rise in fatalities in the country, with around 3,000 people killed in 2015. The Tahrir Institute for Middle East Policy estimated there were more than 100 attacks per month between January and August, compared to 30 per month in 2014. Egypt's security challenges also became more visible on a global scale with the downing of Metrojet 9268, placing the Sinai Peninsula on the map of 'provinces' under the influence of ISIS and suggesting that the group had adopted an increasingly international orientation. However, although the Egyptian security forces have been unable to extinguish the threat, Sinai Province has not made any permanent territorial gains, thereby undermining its claim to being part of the ISIS 'caliphate'.

Nonetheless, disenfranchised young Islamists increasingly fuelled a nationwide insurgency, in the face of the state's violent crackdowns and the absence of any formal mechanisms for political participation. The spread of small but increasingly violent cells and groups is undoubtedly changing the landscape of the conflict and further destabilising the government, which must grapple with emerging threats on multiple fronts. Throughout 2015, Sisi's cult-like following among the population eroded under the pressures of governance and the difficulties of containing intensifying security threats while revitalising the economy. Attempts to stabilise Egypt required a multipronged approach that included improved counter-insurgency operations as well as genuine efforts to encourage political participation and economic inclusiveness.

Iraq

Key statistics	2014	2015
Conflict intensity:	High	High
Fatalities:	18,000	13,000
New IDPs:	2,180,000	
New refugees:	32,000	

The Islamic State (also known as ISIS or ISIL) lost territory in Iraq throughout 2015, as military operations expelled the jihadist group from Tikrit, Baiji, Sinjar and Ramadi. Yet the other actors in the conflict remained highly fragmented, and thus unable to create a powerful anti-ISIS narrative that could transform military victories into political victories. Using their presence on the battlefield to contest power and impose their own visions of society, non-state actors undermined the authority of the government in Baghdad, which accordingly struggled to build an inclusive state, the prerequisite for any potential return to stability.

The shrinking Islamic State

According to Colonel Steve Warren – spokesperson for *Operation Inherent Resolve*, the US mission against ISIS – the group lost 40% of the territory it controlled in Iraq between mid-2014 and late 2015, and 14% of all its territory in 2015 alone. Having been driven out of key areas, ISIS faced an array of challenges, some of which had immediate economic and strategic implications. The loss of territory caused a fall in revenue from taxes and other fines collected from citizens living under the group's rule, weakening its strict bureaucratic structure and restricting funds for further expansion. The group lost lucrative territories, such as Baiji in October, after it switched hands multiple times throughout the year. Baiji was home to Iraq's largest oil refinery until May 2015, when the facility was largely destroyed by ISIS fighters.

The loss of key transportation routes affected the logistical aspects of running the self-declared 'caliphate'. One of the main goals of *Operation Free Sinjar*, launched in November, was to seize Highway 47, which linked the ISIS strongholds of Mosul and Raqqa. The initial severing of the connection between the cities reportedly led to a sharp rise in prices in Mosul. But movement between the two cities did not stop altogether, despite initially diminishing. Within a week, prices had stabilised, as drivers in Mosul found a longer, unpaved alternative route, which ran from Mosul southwest towards Tel Abta and further west to Kirwan (around 20 kilometres from the Kurdish front-lines in Sinjar), then on to the Syrian border.

Nonetheless, the territorial losses projected an image of weakness and controverted the group's narrative of continuous expansion. After its defeat in Ramadi in December, ISIS attempted to reassure its supporters, releasing an audio message purportedly recorded by its leader, Abu Bakr al-Baghdadi. Baghdadi insisted that the 'state' was 'fine and expanding every day', although he acknowledged that there had been 'hardship and strife'. The losses reduced morale among ISIS fighters, and may have induced many of them to desert during the Ramadi operation.

The group's responses to organised counter-insurgency operations remained somewhat consistent, although they resulted in battles that varied significantly in length. Their fighting style was both organised and flexible, relying on rapid movement and frequent, persistent attacks. The majority of ISIS offensives involved simultaneous assaults by multiple suicide bombers driving vehicles, and the detonation of networks of improvised explosive devices (IEDs) in and around cities.

Territorial losses also undermined the state-building aspirations of ISIS, a cornerstone of its project. Even before the defeat in Ramadi, the group displayed frustration with the exodus of Sunni civilians from Anbar, which cast doubt on its claim to be protecting and governing them. In another audio message, Baghdadi called on displaced Sunnis to return to Ramadi and join the caliphate, insisting that ISIS was defending their honour. Similarly, ISIS spokesperson Abu Muhammad al-Adnani appealed to Sunnis facing discrimination in Iraq to join the group. Playing on their fears, he pointed to the 'kidnappings, evictions and killing' of Sunnis in Baghdad, and the fact that the Iraqi government placed heavy restrictions against displaced Sunnis entering most areas of Iraq, including the capital.

Civilians continued to flee from ISIS-controlled territories in Iraq, despite the extreme dangers involved in doing so. Many of those who escaped Hawija were forced to either rely on people smugglers or travel through the front-lines. Furthermore, ISIS sometimes executed civilians who were caught attempting to flee.

The group persevered with its systematic execution of thousands of Iraqis, primarily those who appeared to oppose its rule. Many of the mass killings were part of a broader crackdown that seemed to be motivated by insecurity: a large number of civilians were executed after being accused of espionage or treason, having allegedly cooperated with the Iraqi security forces, the Hashd al-Shaabi (Popular Mobilisation Units, or PMU), the Kurdish Peshmerga or the US-led coalition.

There were frequent reports of ISIS carrying out mass executions, torture and other human-rights abuses in Mosul, as well as of an internal resistance in the city that launched insurgency-style attacks against the group. Although the latter reports were difficult to verify, and it was therefore unclear whether a resistance movement had formed, they suggested that ISIS lacked absolute control over Mosul. Nonetheless, the group was unlikely to be quickly driven out, with the Iraqi security forces' plans to retake Mosul almost certain to encounter further delays and complications. By the end of 2015, the densely populated city had been under ISIS control for one and a half years, and the process of rebuilding the security forces into coherent combat units was far from complete – as had been indicated by their slow progress in Ramadi.

Elsewhere in the country, ISIS had proven resilient. The group's fighters often resurfaced to launch multipronged attacks in areas that it had lost. Key examples of this were several large-scale attacks in Baghdad and Diyala in 2015, as well as retaliatory attacks against the Iraqi security forces after the defeat in Ramadi, and against the Peshmerga after the defeat in Sinjar. ISIS may thus attempt to escalate the conflict in these territories to exacerbate the insecurity on which it thrives. The group's attacks outside Iraq and Syria, particularly in the West, were interpreted by several analysts as a sign of desperation, although others argued that they were an indication of the group's capacity and outreach.

Chaotic battlefields

The military victories against ISIS were complicated by power struggles on several battlefields across Iraq that reflected local, regional and global rivalries. This was discernible in March, with the PMU-led operation in Tikrit; in May, after ISIS took over Ramadi; and in November, in the aftermath of *Operation Free Sinjar*. The offensive in Tikrit, which produced the year's first significant victory against the group, was conducted with the direct involvement of Iran's Islamic Revolutionary Guards Corp (IRGC) and its elite Quds Force – including the latter's well-known commander, General Qassem Suleimani. The US-led coalition did not participate in the first three weeks of the operation, and only intervened following an official request from the Iraqi government. The intervention triggered a backlash from the PMU, in which some militia fighters threatened to withdraw. Ultimately, however, the coalition's aerial campaign proved decisive in the last stages of the battle, and involved tacit cooperation between the parties even as they publicly condemned each other.

These simmering tensions were heightened after ISIS captured Ramadi. The swift takeover, and the collapse of the Iraqi security forces there, resulted in recriminations from all sides. The security forces blamed a shortfall in air support and manpower, while US Secretary of Defense Ashton Carter said they lacked a 'will to fight'. Suleimani characteristically blamed the US-led coalition, arguing that 'only Iran' was participating in the battle against ISIS. The PMU blamed Iraqi Prime Minister Haidar al-Abadi for yielding to US pressure and ordering the prior withdrawal of its fighters, whose presence Washington believed would exacerbate sectarianism in Anbar. One Iraqi commander suggested that former prime minister Nuri al-Maliki was involved in the fall of Ramadi, an event similar to the fall of Mosul in June 2014, when the security forces withdrew. Maliki himself presented the fall of Ramadi as part of a 'Sunni revolution' against Iraq's Shi'ites, implying that residents of the city had collaborated with ISIS – a view that gained credence after the dissemination of a video showing tribal leaders appearing to pledge allegiance to the group. Although the recording had some of the hallmarks of ISIS propaganda, it led some Iraqis to question the loyalties of Ramadi's residents, playing on long-standing mistrust of the Sunnis. Finally, Anbar's Sunni tribesmen insisted that city had been captured because the government had failed to arm them.

These rifts were also evident in plans to retake Ramadi. Even the name of the operation was contentious: after widespread criticism of its initial title, *Labaik ya Hussein* (*We Are at Your Service, Hussein*), chosen in reference to a revered Shia imam, the offensive was rebranded *Labaik ya Iraq*. The PMU and the Iraqi security forces subsequently congregated at Habbaniyah military base – in Anbar, between Ramadi and Fallujah – and began to encircle Ramadi. The PMU, especially Badr Brigade leader Hadi al-Ameri, insisted on retaking Fallujah first. The demand reflected strategic and symbolic calculations linked to Fallujah's proximity to Baghdad and status as the first major city to fall to ISIS (in January 2014). But it was also a deliberate attempt to publicly undercut Abadi by defying his orders.

This resulted in two parallel battles that indirectly competed with each other to reach their objective: the assault on Fallujah, led by the PMU and the Iraqi security forces, and the operation to recapture Ramadi, led by the Iraqi security forces, Sunni tribes and US-trained fighters. Despite benefiting from greater air support and close coordination with US forces, the latter offensive initially stalled. After months of retaking neighbourhoods on Ramadi's outskirts, which ISIS fighters continued to return to, the Iraqi security forces moved into the city in December. The incursion turned on successful US–Iraqi military coordination, and gave the Iraqi security forces a much-needed confidence boost.

There were similar advances in northern Iraq, around the time of *Operation Free Sinjar*. The operation itself lasted less than 48 hours, but it revealed a power struggle between different Kurdish groups, including the Kurdistan Democratic Party (KDP), which dominates the Kurdistan Regional Government (KRG), and several smaller groups, such as the Turkish Kurdistan Workers' Party (PKK) and the Syrian People's Protection Units (YPG). After the PKK and the YPG deployed their forces to several villages near Sinjar, the KDP-linked Peshmerga reversed some of their gains in a bid to control the offensive. Not only was the Peshmerga the strongest of these militias, it also benefited from the US-led coalition's air support and was accompanied by US special forces (although they did not fight on the front-lines).

A speech by KRG President Masoud Barzani in Sinjar immediately after the operation formed part of the group's attempt to strengthen its control over Sinjar. Barzani declared that the district was part of the autonomous Kurdistan region and planted a Kurdish flag in the grain silo in the centre of the city. He also declared, controversially, that no other flag would fly in Sinjar – a statement that heightened tensions with the smaller groups in the area, as well as with Baghdad, since the legal status of Sinjar was disputed. Abadi responded by stating that the Iraqi flag should also fly, pointing to the participation of the Iraqi air force and the Iraqi security forces' assisting role in defusing IEDs.

The KRG expressed an interest in deploying the Peshmerga in other battles against ISIS in northern Iraq, confirming that it was ready to take part in the operations in Mosul, Nineva and Tal Afar. The KRG was likely to use the Peshmerga's participation as leverage over Baghdad, and to demand concessions such as the annexation of disputed territory to a semi-autonomous Kurdish region.

The future of 'liberated' areas

The Iraqi state seemed unable to rebuild and repopulate territories that had ostensibly been secured, and to thereby transform them into a model for a post-ISIS Iraq. In Tikrit and other recaptured population centres, key infrastructure had been destroyed, and the armed groups in de facto control there often carried out human-rights violations – as detailed in a report compiled by the United Nations Assistance Mission for Iraq (UNAMI) and the Office of the High Commissioner for Human Rights.

Several investigations into the PMU's activities in Tikrit suggested that it had violated the laws of war. The PMU's fighters carried out revenge attacks, killing residents with suspected links to ISIS, torching hundreds of homes and looting shops. Abadi attributed these actions to rogue 'gangs' and called for their arrest, but a report by Human Rights Watch suggested that the problem was more deeply rooted. The organisation stated that two large militias, Asa'ib Ahl al-Haq and Kata'ib Hizbullah, were responsible for the destruction of civilian property in Tikrit, and the abduction of more than 200 Sunni civilians, 160 of whom were still thought to be missing at the end of the year.

Abadi denounced the report as inaccurate, and for having exacerbated sectarian strife. His attempt to defuse the situation and divert attention away from the report suggested that he was unwilling to aggravate the powerful PMU, and that he attached great political importance to the success of the military operation. Various Iraqi politicians subsequently insisted that Tikrit had stabilised after ISIS was defeated there, even as independent reports depicted it as a city ruled by rogue militias, with no sign of civilian life.

There were also tensions among competing groups in the aftermath of *Operation Free Sinjar*, as Yazidi civilians carried out revenge attacks on property belonging to Sunni Arabs, looting and/or setting fire to homes and mosques. Many Yazidis vowed to prevent Sunni Arabs from returning to Sinjar. There were also reports of Kurdish fighters expelling Arabs from villages in the area, building on from measures they had taken elsewhere. They reportedly expelled civilians from Sinjar and Zummar in June, claiming that this was a preventative measure designed to clear the area of ISIS sympathisers. While Kurdish forces stated that this only affected around 50 people, other sources stated that hundreds of civilians had been driven out. Kurdish forces were accused of using military victories to alter the demographic make-up of territories contested by Kurds and Arabs, such as the oil-rich Kirkuk province. Many in Baghdad viewed the Kurds as having unilaterally annexed Kirkuk after securing it from ISIS. But several parties, including Baghdad and Shia/Turkmen militias, continued to vie for influence in Kirkuk.

The Peshmerga was only one of many non-state armed groups involved in fighting ISIS that tried to gain leverage over Baghdad, using participation on the battlefield as a means to greater power. Their visions for society were founded on narratives of exclusion that undermined national reconciliation, an essential component of a stable and secure Iraqi state. This failure to provide an inclusive political alternative to ISIS had the potential to benefit the jihadist group, which could exploit the resulting disenfranchisement among Sunnis.

Abadi's weak position

Although Abadi made some progress in curbing the sectarian rhetoric ever present under Maliki, the new prime minister's weakness was reflected in his short-lived attempts at reform. The authority of Abadi's government was increasingly challenged by the array of non-state actors in Iraq, including the KRG and the armed groups that formed the PMU.

During a large-scale protest movement that spread outwards from Baghdad in summer, Abadi seized the opportunity to implement widespread reforms. The demonstrations were spurred by persistent electricity cuts amid an intense heat wave, but they focused on government corruption and mismanagement, especially politicians' alleged theft of the country's resources. Abadi's reforms included the removal of government positions deemed an unnecessary financial burden, such as Iraq's three vice-presidencies, and he received strong backing from Grand Ayatollah Ali al-Sistani. The parliament also reacted positively to the changes, at least initially. Members of the international community praised Abadi and the reforms, portraying them as revolutionary.

But the measures increasingly faced pushback from the Iraqi government, and the country coalesced against him. Having long been a conciliatory figure, Abadi was seen as overstepping his bounds by attempting to go against the system. Many viewed the changes as acts of hostility, particularly after a parliamentary panel blamed the fall of Mosul on 36 government figures, one of them Maliki. The panel argued that the former prime minister had precipitated the army's collapse there by appointing corrupt military commanders, who failed to adequately assess the threat posed by ISIS.

As the momentum of the reforms began to dwindle, Abadi was backed into a corner. His rivals accused him of using the measures to strengthen his control of the government, arguing that the decrees he used to enact some of the changes were unilateral and unconstitutional. Even Abadi's own coalition, State of Law, withdrew its parliamentary support for the reforms, citing the prime minister's failure to conduct sufficiently inclusive consultations.

For Abadi, the reforms were a means to alter the image of his government as fragile and distressed. They were an opportunity to demonstrate his clout, particularly in relation to Maliki, whose political influence persisted. As well as making statements that undermined Abadi, Maliki continued to wield power through his allies. For example, the judiciary, headed by Midhat al-Mahmoud, was closely aligned with Maliki, as were some parliamentary blocs large enough to block legislation.

This resulted in chronic deadlock in parliament, which failed to pass legislation necessary for governing the country, including that to manage oil wealth and amend the constitution. Another source of contention was the National Guard bill, which aimed to decentralise the security forces to local-level recruitment commanded by the governors of provinces. Abadi had pushed for the bill as a means to empower Sunnis, enabling them to participate in the provision of security and thereby reducing the sectarian tensions linked to reliance on the Shia-dominated PMU. This was particularly important in light of the fighting between the various militias of the PMU, which divided loosely into three general factions: groups primarily aligned with Iran, those whose first loyalty was to Sistani and those closely affiliated with Iraqi political parties. Officially, all the militias were under Iraqi command, but in practice many were explicitly sectarian and aligned with Iran.

The PMU's influence was discernible beyond the battleground. Many of its component groups shaped Iraq's political sphere, and some took an explicit stance against Abadi, criticising his close cooperation with Washington as having renewed US power in the country. The pro-Iranian Kata'ib Hizbullah, which formed in 2003 to oppose the US invasion, continuously contested Washington's leadership of the anti-ISIS coalition, and often threatened to attack American troops in Iraq, although this threat was not carried out.

At the same time, the most prominent groups in the PMU viewed Abadi as hostile to their interests, as was demonstrated by a dispute over funding. The PMU demanded in October additional funding for weapons, ammunition and other items, as well as for more than 150,000 additional fighters. It also condemned plans for budget cuts that would affect around 100,000 militiamen who had not registered with the government (but who were promised funding if they did register). The disagreement related more broadly to the organisational command of the militias, with the PMU leadership preferring to receive all the funds and to distribute them at will, while Abadi aimed to use the flow of money to bring the fighters under Baghdad's direct command.

Abadi also faced challenges resulting from the deterioration of Baghdad's relations with the KRG, following the disintegration in mid-2015 of an oil agreement between them. Signed the previous December, the deal enabled Baghdad to sell Kurdish oil, which accounted for around 15% of Iraq's oil sales, in exchange for allocating 17% of the federal budget to the KRG. The KRG cut its supply to Baghdad in June, stating that it had not received its share of funding, and proceeded to sell oil independently – although it later appeared to tone down this bid for economic independence, and agreed to re-enter negotiations with the government. Many in Baghdad viewed the move as an attempt by the KRG to steal Iraqi national resources, and the government threatened to sue any potential buyers of Kurdish oil.

The breakdown of the oil agreement led to severe budgetary constraints for the KRG, which drove it to seize the assets of private banks in areas it controlled. The territory's government was further impaired by internal polarisation over the constitutionality of Barzani serving another term, which exacerbated a split on the ground between the KDP, in Erbil, and the opposition, in Suleimaniya.

Political rivalries between and within local, regional and international entities also continued to undermine governance elsewhere in Iraq, with the result that the country was unable to present a united front against militant violence or focus its efforts on reconstruction and inclusivity. Iraqis were confronted by a multi-layered crisis, of which the immediate security threat from ISIS was just one part. This was reflected in figures published by UNAMI – which recorded the deaths of around 13,000 Iraqis in 2015 (it was likely that there were many more undocumented fatalities) – as well as the International Organization for Migration, which reported that the conflict had displaced 3.2 million people in the year.

As ISIS incurred further losses, one of the main challenges for the Iraqi government was addressing the root causes of the group's success, including the political exclusion of Iraq's Sunni population; the corruption and sectarianism associated with the Maliki-led government; and the fragmentation of the military. The group was unlikely to be eradicated quickly, as the process of driving it out depended on a cohesive military strategy implemented amid an inclusive political atmosphere.

Israel–Palestine

Key statistics	2014	2015
Conflict intensity:	Medium	Medium
Fatalities:	2,500	200
New IDPs:		
New refugees:	2,500	

Prospects for peace between Israel and Palestine remained slim throughout 2015, as their leaderships became increasingly estranged from each other. Against a backdrop of growing domestic frustration with the failures of elite politics, clashes erupted in the West Bank – marking a significant shift in violence away from the Gaza Strip. Renewed tensions surrounding the Temple Mount (known to Muslims as Haram al-Sharif), settlement construction and the Israeli military's use of force were accompanied by a rise in knife attacks and other acts of violence committed by Palestinians against Israelis, particularly towards the end of the year.

Changing theatre of violence

The number of fatalities in Israel–Palestine fell sharply compared to 2014, when around 2,500 people had been killed, most of them Palestinians during the war in Gaza. The ceasefire in Gaza, in place since August 2014, contributed to a massive decline in fatalities: 25 Palestinians were killed in the area in 2015, almost a record low. While there were sporadic clashes on the Gaza border, relative calm prevailed. Even after the collapse in June of the unity government comprising Hamas and the Palestinian Authority (PA), both Israel and Hamas took steps to prevent another escalation of violence in Gaza. Israel also began to ease the blockade on Gaza, with some reports suggesting that the sides were exploring the possibility of indirect talks to reach a long-term settlement.

Both Israel and Hamas shared an interest in preventing the rise of Gaza-based groups affiliated with the Islamic State (also known as ISIS or ISIL), which sought to re-ignite a war with Israel as a means to challenge Hamas's control of the area. The Sheikh Omar Hadid Brigade – named after a man who helped Abu Musab al-Zarqawi establish al-Qaeda in Iraq, the forerunner of ISIS – was believed to have bombed Hamas and the Palestinian Islamic Jihad in July, and to have carried out a series of rocket attacks targeting southern Israel. Although the Israel Defense Forces (IDF) maintained a policy that held Hamas responsible for all attacks emanating from the Strip, Israeli retaliatory strikes targeting the group mostly caused superficial damage to Hamas military infrastructure, as many of the training facilities and buildings targeted were inactive. Some interpreted this as a deliberate move, arguing that any significant attack could weaken Hamas and aid ISIS affiliates who seek to replace them in Gaza. During the year, Hamas conducted a crackdown on these rival organisations.

As the Gaza Strip remained relatively peaceful, the epicentre of the conflict shifted towards the West Bank and East Jerusalem. The latter areas were hit by a wave of violence that grew in the last few months of the year, as a number of Palestinians conducted knife attacks, fire-bombings and vehicular assaults on Israeli troops and civilians. According to data published by the UN Office for the Coordination of Humanitarian Affairs (OCHA), 143 Palestinians were killed by Israeli forces between October and December, compared to 173 for the year as a whole. Furthermore, 133 of these fatalities occurred in the West Bank, nearly three times more than in the same period in 2014.

The escalation of violence in the West Bank had several drivers, including disagreements on access to the Temple Mount, political disenfranchisement, the stalled peace process, continued Israeli settlement activity and heavy-handed security measures. These factors turned Israel–Palestine into a tinderbox, where riots, violent clashes and lone-wolf Palestinian knife attacks could occur at any moment.

In the preceding two decades, right-wing Israelis had pressured predominantly right-wing governments to allow Jews to pray on the Temple Mount, despite the fact that they were prohibited from doing so under a status quo agreement in place since June 1967. Jewish worshippers routinely defied the ban, occasionally provoking a violent reaction from Palestinian groups there. Tensions rose from June onwards, as Muslim and Jewish holidays approached and familiar fears resurfaced concerning access to the Temple Mount. The al-Aqsa Mosque compound, considered the holiest site for Jews and the third-holiest for Muslims, saw clashes between Palestinians, Israeli forces and right-wing Jewish activist groups.

The violence began to increase in September, as Palestinians protested curfews and age-based restrictions on visiting al-Aqsa Mosque. Israeli forces reported finding stockpiles of firebombs, rocks and fireworks at the site, and sporadically stormed the compound, exacerbating Palestinian fears that Israel intended to change the status quo agreement.

Rising aggression

Palestinians in the West Bank increasingly reacted against the strict security policies implemented by the newly formed right-wing Israeli government, which granted its forces wide-ranging powers to curb demonstrations there, as well as stone-throwing, knife attacks and other assaults in the area.

The Israeli Security Cabinet's decision to grant the police more leeway to use their firearms may have contributed to the high number of fatalities and injuries. According to the OCHA, at least 13,589 Palestinians were injured by Israeli forces in the West Bank – 18% fewer than in 2014, mainly during the Gaza war that year. This indicated a trend of rising police violence, especially in the West Bank, where 90% of conflict-related injuries and deaths occurred. According to many reports, the Israeli security services used live ammunition against Palestinian protesters and assailants, prompting human-rights organisations to accuse them of excessive force. Amid the wave of Palestinian knife attacks, Amnesty International stated that there had been a rise in incidents in which Israeli forces deliberately killed Palestinians who posed no threat to life.

According to the IDF, there were at least 100 stabbings, 22 vehicular assaults and 37 shootings in the last three months of the year. In the same period, at least 128 Palestinians and 20 Israelis were killed in clashes between Palestinians and Israeli forces and settlers.

Perhaps the most worrying trend in the violence related to the range of Palestinians involved in the attacks. While most of the assailants were young men, some were 13-year-old boys, teenage girls or elderly men or women – and the majority of attackers were not known to be affiliated with any political faction. The pattern of violence drew some comparisons to the First Intifada, which was disorganised and widely regarded as a people's uprising. While the assaults did not appear to be coordinated, they steadily shifted from Jerusalem to Nablus and Hebron, which as a result were often placed under curfews and other restrictions. Hebron – the unofficial West Bank headquarters of Hamas, a Palestinian city with Israeli settlers living at its centre – was reportedly home to around one-third of the Palestinians who were killed in the attacks.

Settlement construction in the West Bank and East Jerusalem continued to fuel resentment and violence. In October, US Secretary of State John Kerry attributed the increase in assaults carried out by Palestinians to frustration with the unchecked construction of settlements on land that would form part of a

Palestinian state. Such activity had often undermined the Israel–Palestine peace process, including the US-mediated talks that broke down in April 2014.

Another source of frustration was the Israeli authorities' punitive demolition of the homes of Palestinian assailants. The government found itself caught between growing international criticism (all settlements in the Occupied Palestinian Territories were prohibited under international law) and strong opposition to any settlement freeze or reversal from the settlers, as well as the newly elected hardline factions within the ruling coalition. As a consequence, the government sought to placate the latter two groups by providing compensation for any lost settlements. After the Israeli High Court approved the demolition two unsanctioned Israeli housing structures on private Palestinian land in the West Bank settlement of Beit El, the government – which opposed the decision – announced that they would be quickly replaced with 300 new housing units. Yet, despite the government's promise, the court decision sparked clashes between Israeli forces and hundreds of young settlers, who were keen to signal to their government and the international community that they would resist any such moves.

Violence between Palestinians and Israeli settlers in the West Bank continued throughout the year, reportedly causing an average of three injuries and fatalities every two weeks. In one of the most violent retaliatory attacks, the settlers burned down a Palestinian home in the village of Duma, resulting in the deaths of a baby and two members of his family. The incident sparked nationwide unrest that exacerbated discontent surrounding the Temple Mount. In a rare move, Israel's Security Cabinet approved the use of administrative detention to apprehend Israeli suspects in the case. The controversial practice had theretofore predominantly affected Palestinians, who were detained indefinitely without trial on the suspicion of involvement in terrorist activity.

Domestic disillusionment and international criticism

Relations between Israel and the PA reached an all-time low, as the latter continued to lobby for international recognition of Palestine as a state. The effort, which led the PA to join at least 40 international conventions and treaties in 2015, was driven by the lack of progress in US-led peace talks and the perception that the Israeli government was not committed to the two-state solution. Prime Minister Benjamin Netanyahu set out his position on the issue shortly before national elections held in March, when he attempted to rally right-wing voters by promising that a Palestinian state would never come into being under his leadership. While the declaration appeared to stem from a populist scare tactic, it reversed the strong call for a two-state solution Netanyahu had made in a 2009 speech at Bar-Ilan University. Under intense US scrutiny, Netanyahu attempted to backtrack and advocate a two-state solution once again, but Washington remained sceptical, saying it would continue to re-evaluate its approach to supporting Israel in international forums.

Partly due to the inertia of the peace process, Palestinian and Israeli voters increasingly perceived their respective leaderships as ineffective. Many Palestinians had grown wary of President Mahmoud Abbas, and viewed the PA as complicit in Israel's occupation of Palestinian territories. This perception was strengthened by Palestinian security forces' continuing collaboration with their Israeli counterparts in the effort to suppress demonstrations and arrest Hamas members in the West Bank. According to polls conducted by the Palestinian Center for Policy and Survey Research in the second half of the year, two-thirds of Palestinians demanded the resignation of Abbas. Similarly, surveys indicated that the majority of Israelis were dissatisfied with Netanyahu's foreign policy, as well as his handling of the violence. He steadily lost credibility at home after making a failed attempt to prevent the United States, Israel's most significant ally, from helping forge an international deal on the Iranian nuclear programme. His failure drew criticism from the right of the Israeli electorate, while the left largely disapproved of the attempt itself. Furthermore, as the knife attacks continued, many Israelis began to view ultraconservative former foreign minister Avigdor Lieberman as more suited to lead their country.

Due to the absence of moderate elements in the Israeli government, there was little likelihood that the conflict would de-escalate. Having failed to garner support from centrists such as opposition leader Isaac Herzog, Netanyahu maintained his one-seat majority in the Knesset by depending on the ultraconservative, pro-settlement Jewish Home. Meanwhile, Abbas did little to publicly condemn the unrest, prompting the Israeli government to accuse the PA of helping Hamas incite Palestinian violence. Some argued that Abbas's reluctance to do so was aimed at shoring up domestic support, especially among Palestinians who increasingly viewed the PA's security collaboration with Israeli forces as having made them complicit in the occupation of the West Bank and East Jerusalem. Against a backdrop of growing conflict and political intransigence, the majority of Palestinians supported the knife attacks, while many Israelis approved of the extrajudicial killing of those suspected of conducting the assaults.

As the rift widened between the PA and the Israeli government, neither Kerry nor the Quartet on the Middle East were able to bring Netanyahu and Abbas to the negotiating table, despite both leaders indicating that they were willing to meet to quell mounting tensions. The PA's recourse to foreign organisations, some of which seemed increasingly willing to support its cause, generated growing concern in Israel about international isolation. In September, the UN General Assembly overwhelmingly supported a petition to raise a Palestinian flag at the UN headquarters. The same month, the European Union attempted to check the expansion of the settlements by agreeing to label Israeli goods made in the occupied territories. Israel's concerns grew further in December, when US President Barack Obama informed Israeli President Reuven Rivlin that a prolonged stalemate in the peace process would gradually undermine any attempts by Washington to defend Israel in international forums, and to discourage international action in support of the Palestinians.

While the conflict was largely confined to the West Bank in 2015, the violence had the potential to spread to Gaza, where Palestinians faced mounting socio-economic challenges as part of the Israeli blockade – most of which remained in place. Attempts to stabilise the situation seemed to require a sincere effort to reach a joint solution by both the PA and the Israeli government, lest the spectre of a third intifada develop into all-out confrontation.

Lebanon–Hizbullah–Syria

Key statistics	2014	2015
Conflict intensity:	Low	Low
Fatalities:	350	250
New IDPs:		
New refugees:	800	

In 2015 the situation in Lebanon was characterised by mounting instability. On the political level, governmental paralysis triggered unprecedented civil-society protests during the summer. The security environment also deteriorated, as the country experienced its deadliest attacks since the end of the civil war in 1990. Lebanon's frequently tested capacity to adapt to chaotic events enabled it to contain internal and external dangers with relative success, at least in the short term. However, the core issues threatening its long-term stability remained unresolved. There were still fundamental disagreements between the Hizbullah-dominated March 8 alliance, which supports Iran and the Syrian regime, and the March 14 alliance, which is led by the Future Movement and aligned with the Gulf states and the West. This split, combined with the large number of refugees in Lebanon and Hizbullah's intervention in the war in neighbouring Syria, makes Lebanon vulnerable to spillover violence from that conflict.

Political paralysis and civil strife

At the end of the year, Lebanon remained in its longest period without a president, which had begun with the departure of Michel Sleiman from the office in May 2014. In December, parliament failed for the 32nd time to assemble the two-thirds quorum needed to elect his successor.

Despite the agreement that each president would be a Maronite Christian, in place since Lebanon gained independence, the March 8 and March 14 blocs failed to decide on a consensus candidate, and were particularly divided on the issue of the contenders' alignment with foreign powers. Electoral law was also contentious. March 14 candidates supported existing legislation, due to its provisions for maintaining sectarian equilibrium, while March 8 called for its reform in line with proportional representation, which could potentially increase the influence of Shi'ites and Christians. The main contenders for the presidency were Samir Geagea, head of the Lebanese Forces and the nominee favoured by March 14, and Michel Aoun, founder of the Free Patriotic Movement (FPM), Hizbullah's key ally in March 8.

While Prime Minister Tammam Salam took on some presidential duties, Lebanon's basic political mechanisms were debilitated by the crisis. Parliamentary sessions were suspended after they were boycotted by various political factions, especially Hizbullah and the FPM. The refusal to attend parliament was a protest at, inter alia, the lack of negotiations on electoral law; March 14's refusal to consider Aoun as a candidate; and the extension in September of General Jean Kahwaji's term as commander of the Lebanese Army, a role that Aoun wanted to be filled by his son-in-law, General Chamel Roukoz.

Nonetheless, from September onwards, members of Lebanon's political elite took tentative steps towards a political compromise, conducting a national dialogue that involved the leaders of the country's main political forces. Begun by Speaker of the Parliament Nabih Berri, this initiative had made no significant progress by the end of the year, but it prevented tensions from reaching a critical point. The first legislative session of the year was held in November, enabling the country to pass the financial reforms required of it by funding agreements with international organisations such as the World Bank.

Sleiman Frangieh, head of the Marada Movement and initially an outsider in the presidential race, emerged in late November as a viable candidate after meeting with Saad Hariri, head of the Future Movement. The latter party may agree to support Frangieh's candidacy, despite his close ties with Syrian President Bashar al-Assad, in exchange for which Hariri would succeed Salam as prime minister. Given that Frangieh also benefited from Hizbullah's tacit approval, the main obstacle remained a potential veto by Geagea and Aoun, who could justify such a move on the grounds that he had comparatively frail support within Lebanon's Christian community.

The divisions between political groupings affected even basic state services – in ways that deepened the profound sense of crisis in the country, while exacerbating citizens' alienation from, and rejection of, the ruling class. With Lebanon's politicians unable to agree on an area in which to dispose of the country's waste after the long-overdue closure of the Naameh landfill in July, rubbish was left to rot in the streets. Several highly diverse, cross-community activist movements attracting growing numbers of participants led protests against the situation and the political corruption and dysfunction that engendered it. The climax of the protests in late August drew tens of thousands of demonstrators. Violent police repression, arrests and the government's adoption in early September of a plan to move the waste outside major cities brought the protests to an end. Nonetheless, the demonstrations revealed the emergence of an independent civil society that refuses to identify with the existing political and sectarian divisions.

Entanglement with Syria

In early May, Hizbullah launched a large-scale offensive against rebel forces in the Qalamoun region of Syria, which borders Lebanon's eastern Bekaa Valley. The offensive reached its most intense point later that month, as Hizbullah made swift gains with the support of troops from the Syrian regime.

The rapid advances of Hizbullah and regime forces were stalled in early July during their siege of Zabadani, a town near the Lebanese border held by rebel forces. Despite its vast superiority in troop numbers and equipment, Hizbullah was forced in September to accept a six-month ceasefire due to fierce resistance. The agreement allowed for the safe departure of rebel troops from Zabadani. Hizbullah's offensive brought the greater part of the Lebanon–Syria border area under its control, buttressing the group's stronghold in the Bekaa Valley and Lebanon as a whole against rebel incursions. However, Hizbullah sustained heavy troop losses and faced mounting criticism at home, leaving it exhausted. Russia's aerial campaign to support the Syrian regime, begun in September, therefore provided respite for the group at a critical juncture.

Jihadist and other rebel forces, particularly al-Qaeda affiliate Jabhat al-Nusra, retained a strong presence near Arsal, a Sunni-dominated Lebanese town near the northeastern border with Syria. The Lebanese Army also reinforced its deterrence posture in the area, and occasionally engaged in deadly clashes with rebel groups. But despite recurrent attacks, such as a November car bombing that killed at least seven people in Arsal, the army declined to conduct a full-scale offensive to secure the town, deeming the operation too dangerous given the strength of rebel forces in the area and their pockets of support among residents. Lebanon also wanted to avoid an open confrontation with the rebels, lest this bring Syria's increasingly sectarian conflict deeper within its own borders. While Hizbullah consistently advocated a more hawkish stance, the group left the matter to the army due to the confessional make-up of the area.

Tripoli, the largest Sunni city on Lebanon's north coast, remained a major hub for jihadist activity. The most significant attack there occurred in January, when two suicide bombers killed nine people and injured 37 in Jabal Mohsen, a pro-Assad Alawite neighbourhood of the city. Numerous police and military operations were carried out across the country throughout the year, foiling two terrorist attacks in Tripoli in November, and leading to the arrest in August of Sheikh Ahmad al-Assir, a radical Salafist who had been in hiding for over two years.

However, the security forces were unable to prevent two suicide bombings in Burj Al-Barajneh, a Shia neighbourhood in southern Beirut, on the same day that the planned attacks in Tripoli were halted. Killing 43 people, the terrorist attack was the deadliest in Lebanon since the end of Lebanon's civil war, and the first for which the Islamic State (also known as ISIS or ISIL) claimed responsibility. The jihadist group targeted the Hizbullah-controlled area in retaliation for the latter's support of Assad. The Lebanese authorities feared that, having displayed both the means and the will to expand its operations far beyond Syria and Iraq throughout 2015, ISIS would use the attack as a starting point for an enduring campaign to destabilise Lebanon. Lebanese security sources also suggested that ISIS networks were far more resilient and difficult to uncover than those of rival jihadist groups, significantly increasing the threat they posed.

Refugee crisis

Lebanon remained the country with the highest number of refugees per capita. According to figures published in May 2015 by the United Nations High Commissioner for Refugees (UNHCR), around one-quarter of its population were refugees, approximately 1.1 million of whom were Syrian (in unofficial estimates, the actual number was closer to 1.5m). From that point on, the UNHCR obeyed the Lebanese government's instructions to stop registering new refugees, in an effort to limit their presence in the country. Thus, new Syrian refugees arriving in Lebanon did not benefit from refugee status and the humanitarian assistance it guaranteed, which contributed to the deterioration of their living conditions. The refugees posed a number of challenges to Lebanon. In terms of security, terrorist groups were able to use their affiliations with some refugees to aid them in conducting operations in the country. In its statement claiming responsibility for the Beirut bombings, ISIS revealed that the attackers were two Palestinians and one Syrian, and several Syrians were arrested in the subsequent police operations. Lebanon's Education Minister Elias Bou Saab even said in September that one in every 50 refugees entering Europe could be a member of ISIS. Though undoubtedly an exaggeration, this statement was highly indicative of the widespread perception of Syrians as having brought the conflict into Lebanon. Moreover, Lebanon's fragile sectarian make-up relied on a tenuous demographic equilibrium between religious communities. According to many Lebanese, the vast majority of Sunnis among Syrian refugees threatened this equilibrium, as did the influx of Palestinian refugees before them. Such sentiment and the economic problems created by the refugee crisis (such as rising unemployment and public expenditure) led to tensions with the local population, and at times even violence. The refugees were also victims of numerous human-rights abuses, including extrajudicial detention in squalid conditions that fostered their sense of alienation.

Yet Lebanon coped with the refugee crisis and largely contained the threats it faced. International aid was important to these tasks, through both financial assistance and the work of organisations such as the UNHCR, with the Third

International Humanitarian Pledging Conference for Syria promising US$450m to Beirut in March. However, this aid decreased throughout the year, and Lebanon received far less than the US$2.14 billion it had requested. Assisted by the UN, Lebanon asked in December for an additional US$2.48bn to finance the Lebanon Crisis Response Plan for 2016. Fatigue among donors, exacerbated by their frustration with Lebanon's inefficient political system, could be the cause of this shortfall in aid. But the increasing number of refugees, four out of five of whom are living in poverty, means that financial assistance is more crucial than ever to the country's stability.

The combined effect of the unresolved political and security challenges was to bring the country closer to the conditions that caused civil war to break out there 40 years earlier. These conditions included significant regional divisions that affected domestic politics, a sclerotic political system unable to deal with widespread communal grievances, and deepening tensions between Lebanese and refugees. Comprehensively addressing these problems, rather than simply devising a short-term containment strategy, was vital to avoiding descent into even more profound turmoil and disintegration.

Libya

Key statistics	2014	2015
Conflict intensity:	High	High
Fatalities:	3,000	2,000
New IDPs:	340,000	
New refugees:	1,250	

The crisis associated with the civil war in Libya deepened considerably in 2015. A bloody power struggle between two rival factions was a major cause of the fighting that occurred across the country, exacerbating an already dire humanitarian situation and further raising the casualty rate. In a bid to gain legitimacy, the country's two governments and their respective coalitions continued to fight for political and territorial control. Violent clashes, expanding front-lines and serious violations of international human-rights law contributed to an increasingly unstable security landscape. Amid international concerns that the absence of a functioning state was turning Libya into a stronghold for terrorist organisations, the spotlight was on a national peace process designed to form a unity government and to address the underlying causes of the conflict.

Stalled negotiations

The year began with a fresh round of UN-sponsored talks in Geneva between the Tobruk-based House of Representatives (HoR) and the Tripoli-based General National Congress (GNC). The GNC is supported by Libya Dawn, a coalition of Islamist groups, hardline revolutionaries, Misratan brigades and tribes from western Libya. The HoR and government led by Prime Minister Abdullah al-Thinni is backed by the Libyan National Army (LNA), secular and nationalist factions, Zintani-based militias and former regime elements. With hostilities and tensions running high, UN Special Envoy Bernardino León relied heavily on shuttle diplomacy during the first half of the year to navigate the reconciliation process.

Since its establishment in March 2014, the United Nations Support Mission in Libya (UNSMIL) had always faced a difficult task. The peace process proved to be gruelling and León, who headed the mission, had to manage a negotiation unbalanced by the fact that the HoR was internationally recognised while the Islamist-leaning GNC was not. His initial approach appeared to follow a loose, top-down model in which high-level, centralised engagement, backed by international stakeholders, would lead to executive decisions. Subsequent engagement with actors on the ground would be needed to implement policies. But this approach did not fully account for the power dynamics in Libya, especially the important role played by tribal and regional groupings in forming the country's social fabric. Mediation in which tribal leaders and municipal councils act as local power brokers is a widespread practice there. Despite later efforts by the resource-limited mission to host workshops with these actors, complaints of underrepresentation were regularly voiced. Another key obstacle to the negotiation process was that, at critical junctures, any move by the United Nations was interpreted by one side as unduly favouring the other. Accordingly, as the UN tried to placate one coalition, the other camp would criticise the move. León could only succeed by striking the correct balance – an undertaking that became increasingly difficult as fighting on the ground grew in both scale and complexity.

One of the most significant points of contention between the HoR and the GNC was the role of the State Council, designed to be the country's second chamber until a new constitution comes into effect. It would be mostly be composed of current GNC members. The GNC did not accept that the State Council was a purely consultative body with no legislative powers. In an attempt to address this perception, León made a key revision to Article 19 of the proposed political agreement in June, granting the State Council the power to express 'binding opinions with a qualified majority on legislation', which the planned unity government would then present to the HoR. However, as the accord also provided the GNC with 90 of the 120 seats making up the State Council, representatives from the parliament based in Tobruk argued that they would only be able to pass laws that had already been vetted by the State Council and, by extension, the GNC. As a result, the HoR rejected the draft and demanded more balanced representation, calling for the State Council to have reduced responsibilities and to be relocated from Tripoli, the proposed headquarters of the unity government, to the town of Sabha, in southern Libya. This forced the UN to go back to the drawing board. Eventually, in an effort to mollify the HoR, León proposed that the 90 seats held for the GNC would only be given to representatives originally elected in July 2012. This move, in turn, was met with considerable resistance from Islamist hardliners in Tripoli.

Territorial gains by ISIS

A major change in Libya's increasingly volatile security landscape was the increase in activity by new players, particularly jihadist group the Islamic State (also known as ISIS or ISIL). Having announced in 2014 the creation of three 'provinces' in Libya – demonstrating an aspiration to establish a major foothold outside of Iraq and Syria – the group capitalised on the pervasive disorder to make significant territorial gains. It also carried out a number of deadly attacks in 2015, causing many civilian casualties. Although the Libyan branch of ISIS did not have the operational or financial capacity of its Iraqi and Syrian counterparts, the group appeared to view Libya as an operational zone for its fighters, a haven from front-line fighting and potentially a fallback capital. A UN monitoring committee estimated that ISIS accumulated around 2,000–3,000 fighters in Libya during the year; approximately 800 of them were believed to have previously fought in Syria and Iraq.

Headquartered in Sirte and Derna, the Libyan branch of ISIS also had significant influence in Tripoli, Benghazi and Ajdabiya. It increasingly relied on the brutal tactics for which the group was notorious. One example of this came in mid-February, when it published a video showing the beheading of 21 Egyptian Coptic Christians in Libya. This sparked alarm among the country's neighbours, as well as governments further afield with vested interests in the security of the country. Egypt retaliated with force, conducting airstrikes in Derna and Sirte that killed ISIS fighters and destroyed ammunition stockpiles. Undeterred, ISIS subsequently demonstrated the breadth of its reach and capacity to conduct further attacks, engaging in armed clashes across the country and targeting diplomats and expatriate residents.

Having pledged allegiance to ISIS leader Abu Bakr al-Baghdadi in October 2014, the powerful Islamic Youth Shura Council continued its work to establish an Islamic caliphate in Libya. This helped ISIS assert control over Derna, although rival Islamist groups maintained limited influence in some districts, as well as in areas on the outskirts of the city. In the absence of a functioning state, ISIS created a public administration and enforced its strict interpretation of sharia law, which resulted in summary executions, public floggings and patrols by religious police. The Libyan National Army conducted airstrikes and offensive ground operations, but ISIS managed to maintain control of Derna for eight months. In June 2015, members of ISIS assassinated Nasser al-Aker, leader of the Derna-based Abu Salim Martyrs Brigade, after he refused to pledge allegiance to their group. The Mujahedeen Shura Council of Derna, an ally of Aker's organisation, declared that it would wage jihad on ISIS 'until none of them are left'. A week of violent clashes ensued, until fighters from the Mujahedeen Shura Council of Derna and armed residents expelled ISIS from the city. The defeat reflected a failure by ISIS to cultivate support among locals or to create partnerships with Derna's other Islamist factions.

Around the same time, ISIS seized control of Sirte, along with 200 kilometres of coastline east of the city. The group had first arrived there in February, and began slowly making territorial gains while the GNC-aligned Misrata Brigades, led by Brigade 166, was preoccupied with fighting the Libyan National Army for control of Sider. After occupying key infrastructure and setting up checkpoints in Sirte, ISIS engaged in a major battle with Brigade 166 near al-Ghardabiya air base, eventually forcing the latter to withdraw completely from the city. Brigade 166 partially blamed the GNC and its affiliated Libya Dawn coalition for providing it with little support, soldiers or weapons. Despite facing resistance from locals, ISIS quickly raised its black banners around Sirte and attempted to reproduce the bureaucratic structure it had established in Raqqa.

Although the group faced a similar sequence of events as those in Derna, the outcome was more bloody. In August, ISIS killed conservative imam Khalid bin Rajab Ferjani after he refused to relinquish control of his mosque, sparking an uprising by members of the Ferjani tribe and local armed groups, which attempted to drive out ISIS. The HoR's ambassador to France, Chibani Abuhamoud, declared that between 150 and 200 people were massacred in the ensuing battles between the sides. Perhaps learning from its experience in Derna, ISIS brought in reinforcements to conduct a brutal crackdown. Resistance leaders were beheaded and their bodies crucified and strung up in public. This put an end to the insurrection, and allowed the group to maintain control of Sirte.

On 21 September, as territorial gains made by ISIS exacerbated tensions, León announced that the UN had completed its work. He circulated what he claimed to be the 'final text' of a peace deal and invited the negotiation teams to submit their nominations for the Presidency Council. While the HoR made recommendations, the GNC postponed doing so. Nonetheless, in a bid to move the process along, León announced the recommended composition in October, stressing that all the names he introduced were merely suggestions. The plan stipulated that Faiez Serraj, a legislator from the HoR, would be prime minister, and that there would be three deputy prime ministers – one from the west, another from the east and the last from the south. Two senior ministers would complete the council.

The reaction to the announcements varied greatly across the country, and the peace process entered a critical stage: UNSMIL needed to convince all parties to forgo further amendments to the text and sign the political agreement, thereby equipping the unity government with the tools and support it needed to combat domestic terrorism. The formation and global recognition of a unity government would be an important step towards lifting the international arms embargo in place since 2011. Despite receiving the backing of the international community, the peace process was jeopardised by a scandal involving León and fragmentation within each coalition, both of which contributed to the delay in establishing a unity government. The scandal broke in November, following the publication of emails from high-level UAE officials that suggested León favoured the HoR during negotiations, a position allegedly linked to his acceptance of a job as director general of the Emirates Diplomatic Academy. Under UNSMIL's plan, León would negotiate the peace process and his successor, Martin Kobler, would oversee its implementation. Kobler took up his role on 17 November, and immediately met with the different factions in an attempt to remedy the situation. Although he faced pressures to amend the agreement in light of the allegations levelled at León, Kobler stressed that this was no longer possible.

International support for a unity government

Convincing the GNC and HoR to sign the political agreement proved to be complicated. At the beginning of the peace process, the parameters of the Libyan battlefield seemed simple: two rival governments and parliaments, each supported by a multitude of militias that were seemingly united in their ideas. Since then, there had emerged a series of divides along ideological, regional and tribal lines within both Libya Dawn and the forces supporting the HoR. This fragmentation contributed to escalating tensions between the groups and was instrumental in shaping the peace process. Hardliners in both camps used the loss of faith in the UN as an avenue to convince moderate factions that more extreme action was necessary. This impeded efforts by the UN to establish a unity government.

The risks became apparent in early December, when representatives of the GNC and the HoR met in Tunisia without the involvement of the UN. This led to the signing of the Tunis Agreement by Awad Abdul Saddeq, GNC first deputy president, and Ibrahim Fethi Amish, head of the HoR's National Dialogue Committee. The deal was designed to establish two committees – one to name a prime minister and two deputies, and the other to review the constitution. While the effort initially drew tentative praise in the foreign media, it soon became apparent that the agreement was not backed by a general consensus in either coalition. In Tobruk, 92 members of the HoR formally criticised the deal, claiming that Amish had no authority to take part in negotiations, and that the undertaking was an attempt by hardliners in their parliament to derail the UN process. The GNC's leadership welcomed the Tunis Agreement, but discontent in the wider parliament suggested that this sentiment was not shared by all.

With Kobler calling for the swift implementation of the UN agreement, HoR President Agila Saleh and GNC President Nuri Abu Sahmain met on 15 December in Malta. There, they rejected the deal on the grounds that, as Saleh put it, 'we are able to work our problems by ourselves'. Yet, on 17 December, HoR Deputy Speaker Emhemed Shoaib and GNC Deputy President Salah al-Makhzoum signed the UN agreement in the presence of 27 GNC members and 80 HoR representatives. Although both Saleh and Abu Sahmain denounced the move, arguing that Shoaib and Makhzoum did not have the authority to sign the deal, the UN Security Council quickly moved to adopt Resolution 2259, recognising the unity government headed by Faiez Serraj as the sole legitimate leadership of Libya.

The Security Council's vigour stemmed from a widely held interest in establishing a strong, sustainable unity government in Libya. As well as being an arena for a bewildering array of armed actors, the country was to some extent a battleground for a proxy conflict between outside powers, as each coalition was supported by states that shared its agenda. The anti-Islamist United Arab Emirates and Egypt backed the HoR. Qatar and Turkey favoured the GNC, largely due to their support for the Libyan branch of the Muslim Brotherhood, the Freedom and Construction Party. These countries have not only hosted talks and visits in their capitals, but have also been accused of shipping weapons and supplies to both camps. However, the growing cross-border threat posed by the operations of the local branch of ISIS may have prompted these interested parties to reassess their activities in Libya.

The threat was perhaps most apparent in Tunisia, which was struck by a series of deadly terrorist attacks. Responding to these incidents, Tunisian Secretary of State for Security Rafik Chelly suggested that 'everything is being planned in Libya', and that the senior commanders of Tunisian terrorist groups were primarily based in the country. With the apparent proliferation of terrorist training camps in Libya and regular arrests of recruiters for ISIS operating in Tunisia, the government announced in January 2015 that 3,000 Tunisians had travelled abroad to fight for the group. After 13 people were killed in a November attack on a bus carrying members of the Presidential Guard, Tunisian President Beji Caid Essebsi put in place stringent security measures, declaring a state of emergency, closing the borders with Libya and approving the recruitment of 6,000 more officers by the national security forces.

The European Union also closely followed developments in Libya. Due to its proximity to the country, Italy was particularly invested in the peace process there, sending Minister of Foreign Affairs Paolo Gentiloni to co-chair a meeting with US Secretary of State John Kerry, promoting the UN negotiations and promising financial and security support for the unity government. The EU hoped that the success of the Libyan peace process would stem the flow of migrants from the Middle East to Europe, a crisis of growing concern to all its member states.

Immediate challenges and sources of instability

The rising instability in Libya significantly exacerbated the country's humanitarian crisis. As regular clashes between the rival coalitions and ISIS led to serious violations of human rights, Libyans and resident foreigners were displaced on a massive scale. The UN High Commissioner for Refugees reported on 30 June that the number of internally displaced persons in Libya had doubled since September of the previous year, to more than 434,000. Furthermore, increasing threats to civilians prompted thousands of people to risk their lives by attempting to cross the Mediterranean aboard boats and dinghies. Despite a series of naval rescue missions by EU countries, the International Organization for Migration reported the deaths of 3,771 people in 2015, with 77% of these occurring in a route mostly used by human traffickers from Libyan shores. In a particularly tragic incident in August, an estimated 800 people drowned when the vessel carrying them sank off the coast of Tripoli.

The unity government therefore faced many intimidating challenges – not least, the risk that it might become merely the focal point for a third coalition. As Serraj began to appoint the members of his cabinet, international pressure was mounting to formulate a plan for countering ISIS in Libya. The latter task was perhaps made more difficult by the group's November attack on Paris, which led to the intensification of Western airstrikes on ISIS positions in Syria and Iraq. The group's leadership subsequently attempted to tighten its grasp on Sirte, possibly viewing the city as an alternative base of operations. This idea was bolstered by the mantra 'Sirte will be no less than Raqqa', frequently repeated in ISIS sermons and radio broadcasts.

Nonetheless, ISIS struggled to gain popular support in Libya. With an array of different factions on the ground, ISIS had an inherent disadvantage as the perceived outsider among local groups, despite largely consisting of Libyan fighters who had returned from the conflicts in Syria and Iraq. This could be seen in the significant resistance and even outright rejection that ISIS met with in both Derna and Sirte. Coupled with a relative lack of exploitable sectarian divisions in Libya, the group faced considerable obstacles in its mission to further expand in the country. Thus, Serraj and the unity government had an immediate opportunity to exploit this weakness by implementing measures on the ground to make Libyans and foreigners feel safer. This investment will be critical in winning public support.

Mali (The Sahel)

Key statistics	2014	2015
Conflict intensity:	Medium	Medium
Fatalities:	500	350
New IDPs:	19,000	
New refugees:	16,000	

The peace process undertaken by Bamako and Mali's insurgent groups made limited advances in 2015, against a backdrop of deepening social tensions. While several initiatives helped de-escalate the conflict between the sides, it was unclear whether this progress could be sustained. This was partly because the peace agreement somewhat pragmatically rested upon perpetuating the northern politico-military status quo, secured with the promise of additional national and international funds to support the north. The status quo is tied to illicit trade and criminal activity and therefore inimical to the good governance necessary for sustaining peace. Moreover, the country's security continues to be undermined by a number of Islamist militant groups. These factions are largely located in northern Mali, but their violent activities are increasingly felt in the centre and south, including Bamako. The persistence of such terrorist networks, in spite of French military intervention, UN peacekeeping forces and strengthened regional security cooperation, became an additional concern for 2015. Challenges in 2015 included implementing the Algiers Accord and addressing the violent threat posed by these jihadists, who remain politically and ideologically outside the formal peace process.

Political compromises

An official peace agreement known as the Algiers Accord was signed in June 2015. To reach this deal, compromises were made by the government and the two main insurgent groups based in the north of the country, the relatively pro-government Platform and the more secessionist Coordination des Mouvements de l'Azawad. A major part of this was the Coordination's abandonment of its goal of establishing an independent territory, and its agreement to back a unified, secular Malian state. Bamako, for its part, agreed to make 'Azawad' the official name of the country's three northern regions. Significantly, the Algiers Accord devolved considerable power to the Tuareg communities represented by the Coordination. It also promised economic investment in the north, as desired by both the Platform and the Coordination. Meanwhile, both groups agreed to a process for their disarmament, demobilisation, reintegration and reinsertion – through measures that included the incorporation of some of their units into national security forces. The establishment of the Truth, Justice and Reconciliation Committee, which aims to investigate violence and other human-rights abuses committed in the country in 1960–2013, is an important step towards fostering a culture of reconciliation.

However, several key developments suggested that lasting communal reconciliation remained a distant goal. Firstly, the Coordination initially refused to sign the peace accord, arguing that the deal contained inadequate security arrangements and ignored some of the group's aspirations. The move was almost certainly prompted by concerns that its forces would not be stationed optimally on strategically important northern roads, checkpoints on which would offer local political and economic benefits. Secondly, both the Platform and the Coordination broke the ceasefire agreement central to the accord several times between May and September, halting the implementation of the deal. Each breach was a result of insurgent units moving into northern territories forbidden under the terms of the peace agreement. Their objective was to gain control of the key northern roads before being tasked with policing them after the accord was implemented. The last of several battles caused by this manoeuvring took place on 17 September, and resulted in 15 fatalities. It was followed by a meeting between the Platform and the Coordination in Anefis, where they agreed on 15 October to end the hostilities.

It is likely that this 'pact of honour', as they called it, was contingent on the pledges of significant financial support announced shortly thereafter. Building on the formal accord with the government, the unofficial deal between the two northern groups met with a positive international response. France declared on 21 October that it would provide Mali with reconstruction assistance of €360 million, €80m of which would be spent on the north. The following day, the Organisation for Economic Cooperation and Development (OECD) hosted an international conference on Mali, aiming to solicit further investment from the public and private sectors. The participants in the meeting concluded that Mali would need around €3.5 billion in humanitarian and development assistance in the following six years.

Economic and ideological drivers of conflict

The insurgents' deep concerns about controlling roads in the north relates to the smuggling prevalent in the area. Since the 1970s, Mali's northern border has been a transit point for a growing trade in, inter alia, people, cigarettes, narcotics and arms. Many northern communities depend on this illicit commerce, and require control of roads between Algeria and Mali to collect bribes and other kickbacks. The region also continues to experience kidnappings, particularly by terrorist groups such as al-Qaeda in the Islamic Maghreb (AQIM) seeking to collect ransoms. Analysts point to a degree of collusion between trafficking networks and terrorist groups intent on profiting from kidnappings. The links exist even within some terrorist groups, such as al-Mourabitoun (led by Mokhtar

Belmokhtar), which began to emerge through patrimony generated by illicit smuggling.

The fact that so much of the peace process has, in effect, hinged on control over trafficking routes has two key implications. Firstly, corruption and other criminality is endemic in the north, and reportedly in Bamako, undermining Mali's long-term prospects for security and good governance. Secondly, such activity is often motivated by a desire to acquire revenues for the deprived communities of the north, where economic marginalisation is pervasive. The region is mired in the kind of poverty that constitutes a major driver of conflict. Thus, Mali's deep-rooted structural problems suggest that the insurgency will persist, despite efforts to achieve peace. For the conflict to be brought to a permanent end, criminality in the north needs to be tackled by force, as well as by improving governance and strengthening the rule of law. There needs to be broader recognition of the problem among the Malian authorities, which continue to largely ignore cross-border trafficking, instead identifying radical Islamists or tribal dynamics as the key drivers of the conflict.

The international community must confront corruption among political elites in Bamako, and support the Malian government in creating legal forms of employment that could replace the criminal economy. In this way, the northern insurgent groups would be legitimately empowered to negotiate peace with Bamako. France's pledge of aid was undoubtedly important, but the promised funds would be delivered over a period of years, and it would take far more than €80m to truly bolster the north. The Malian government, meanwhile, needed to honour its commitments in the Algiers Accord to integrate the north politically and economically.

Unpredictable attacks on civilian and military targets occurred throughout the year. The assault on the Radisson Blu hotel in Bamako on 20 November was a reflection of the complexity of Mali's protracted conflict, in which, despite progress between the official warring parties, a number of violent jihadist factions remain politically and ideologically outside the peace agreement. The resulting one-day siege led to the deaths of 20 people, including the two perpetrators, jihadists from radical Islamist group al-Mourabitoun. While in many attacks the assailants were not officially identified and no armed group claimed to have been involved, there were often indications that Islamist militants had been responsible. For example, an assault carried out by unidentified gunmen near Timbuktu on 17 December, in which three men were killed, occurred outside a Christian radio station, suggesting that the perpetrators had a religious motive.

Another indication that jihadist violence was spreading south was provided by the Macina Liberation Front (FLM), an Islamist militant group based in the centre of the country and largely composed of ethnic Peuls. On 13 August, the FLM shot and killed Imam Elhadji Sekou Ba, who had publicly discouraged Muslims from taking up arms. Yet unlike the northern jihadist networks focused on the imposition of and adherence to sharia law and jihad against the French military and UN peacekeeping mission, the FLM's rhetorical focus was more on the inequality and corruption of the state, a message that had the potential to gain traction in marginalised communities.

Having concluded *Operation Serval* in July 2014, the French military's *Operation Barkhane* continued to weaken jihadist groups in northern Mali, including al-Mourabitoun, Ansar al-Dine and AQIM. France reported in late 2015 that *Operation Barkhane*, a pan-Sahel effort, had led to the death or capture of 125 terrorists and the seizure of 20 tonnes of munitions in Mali. Nonetheless, there was a possibility that this progress was only temporary, given the operation's costs and reliance upon the continued political commitment of 'G5 Sahel' countries: Burkina Faso, Chad, Mali, Mauritania and Niger. The UN Multidimensional Integrated Stabilisation Mission in Mali (MINUSMA) also helped stabilise the conflict, despite having the fourth-highest casualty rate of any UN peacekeeping force ever. Furthermore, Mali's neighbours were increasingly coordinating their security operations as part of the Nouakchott Process, which started in late 2014 and included Algeria, Guinea, Côte d'Ivoire and Libya, as well as the G5 Sahel. This process should yield results in promoting coordinated security operations and intelligence sharing but since it remains in its infancy, it is difficult to fully assess its impact.

Persistent humanitarian problems

In many parts of Mali, civilians continued to be affected by banditry, inter-communal violence, indiscriminate jihadist attacks and retaliation by militants. They were often harassed or forcibly displaced, and some communities were denied freedom of movement, particularly due to skirmishes between warring groups in the north. This could be seen in June in Menaka, where fighting between the Coordination and the Self-Defence Group of Imrad Tuareg and Allies (GATIA), a Platform affiliate, led to the harassment and displacement of locals.

The threat from inter-communal conflict was evident in in July in Talataye. MINUSMA reported that there had been an escalation of a dispute over cattle theft in the village, centring on members of the Daoussak and Imghad Tuareg communities, who were aligned with the Coordination and the Platform respectively. On 18 July, a land dispute sparked violent clashes between Peul and Dogon communities, leading to the deaths of six people. Two days later, there were more clashes between the Imghad and Daoussak communities in Indelimane, resulting in the execution of two Daoussaks.

There were also frequent reports of human-rights abuses carried out by Malian security forces, the Coordination, the Platform, jihadist groups and other armed organisations. Between 12 June and 1 August, MINUSMA recorded two sexual assaults – one in Timbuktu region, committed by a Malian soldier against a minor, and another in Gao region, carried out by two members of a Coordination affiliate. A dozen children have been identified as fighting with armed groups: eight with GATIA in Menaka; three with Movement for Unity and Jihad in West Africa in Mopti; and one with Ansar al-Dine in Kidal region.

Such insecurity often prevents humanitarian assistance from reaching civilians, particularly in Gao and Timbuktu regions, and aid organisations have had to rely on aircraft to support some of their operations in northern Mali. Furthermore, there were at least 60 security incidents involving humanitarian personnel in 2015, three times more than the previous year. At least 100tn of food aid was lost to looting in 2015. The scale of the problems faced by civilians in Mali was reflected by the UN's humanitarian appeal for the country, which at US$377m for 2015 was the second largest among the nine Sahel states. Yet by the end of December, the UN had received only 35% of the necessary funding.

Mali's food security remained fragile. In late November, the UN reported that more than 2m Malians experienced food insecurity (defined as inadequate food access, availability, utilisation and stability) and that more than 400,000 suffered from severe food insecurity (defined as not consuming enough calories and micro-nutrients each day to maintain their health). In Timbuktu region, more than 17% of children under the age of five were reportedly suffering from acute malnutrition. The number of people supported by the UN High Commissioner for Refugees – including asylum seekers, refugees, returnee arrivals during the year (former refugees) and internally displaced persons – dropped from 244,000 in January to 157,000 in December. In many areas, schools that had closed in 2012–13 were yet to reopen. These included 94 schools in Mopti region, 86 in Timbuktu region, 53 in Kidal region and 52 in Gao region. Compounding the challenges faced by civilians, there was a measles outbreak in Kidal region in April, which Mali's government, the UN and other organisations contained. At least one health threat was negated in 2015, however, with the apparent eradication of Ebola in the country.

Despite persistent humanitarian problems, there was a marked improvement in the overall situation in Mali during the year. Genuine progress in the political process, with the acceptance of formal and informal peace agreements, led to a considerable decrease in violent attacks – albeit while having little effect on the underlying economic drivers of the conflict. One of the most significant threats to Mali's stability remained Islamist militant groups, which continued to operate out of the north of the country despite the efforts of the French military, MINUSMA and neighbouring states to contain and defeat them. These groups have repeatedly proven their ability to carry out large, complex attacks on civilian targets in the south, as al-Mourabitoun did on 20 November in Bamako. Mali's human security remained weak, with civilians often affected by the activities of armed groups, inter-communal clashes and high levels of poverty and underdevelopment. Implementing the Algiers Accord was essential to truly addressing the causes of the conflict. A critical part of this process would be strengthening governance in Bamako and less populous areas, particularly Timbuktu, Gao and Kidal regions.

Syria

Key statistics	2014	2015
Conflict intensity:	High	High
Fatalities:	70,000	55,000
New IDPs:	1,100,000	
New refugees:	1,650,000	

The conflict in Syria entered its fifth year in 2015. The opposition to President Bashar al-Assad coalesced into stable and successful alliances for the first time, having been plagued by infighting and poor coordination between groups with wildly different ideologies. Islamist rebels emerged as the pre-eminent force in northwestern Syria; moderate and secular forces gained control in the south; and, in the north-east, Kurdish and moderate Arab fighters rolled back the advances made the previous year by the Islamic State (also known as ISIS or ISIL). However, ISIS continued to control most of the east. The regime's territory shrank in the first half of 2015, and it became ever more dependent on military support from the Lebanese militia Hizbullah, as well as Iranian military and paramilitary forces. By late summer, the military balance appeared to have tipped in favour of the opposition.

Foreign forces became ever more entangled in the conflict. Canada, Turkey, Australia, France and the United Kingdom all joined the US-led coalition conducting airstrikes against ISIS in Syria. In September, Russian warplanes also entered Syria's skies, providing direct military support to the Assad regime. Moscow claimed to be bombing only terrorist groups but targeted rebels of all stripes, including factions supported by the United States, along the regime's most vulnerable front-lines. The intervention allowed the regime to regain the initiative, thereby escalating the conflict. Meanwhile, foreign diplomats pursued a peace deal with renewed impetus and, after a series of conferences in Vienna, they arranged for negotiations between the regime and the opposition to begin in January 2016.

Rebel stability

The rebels organised into five main blocs, each with their own territorial concerns. Jaysh al-Fateh, one of these alliances, was formed in March by jihadist and conservative Islamist factions, including Ahrar al-Sham, the largest rebel group, and al-Qaeda affiliate Jabhat al-Nusra. Jaysh al-Fateh quickly became the dominant military force in the northwest: within four days of its foundation, it had driven regime forces out of the city of Idlib – the first provincial capital to fall outright to the opposition since the beginning of the war. In the following two months, the rest of the province fell, with the exception of the Shia-majority towns of Kefraya and Foua. The alliance then sought to expand its territory further south, into Sahl al-Ghab in northern Hama Province, but it made only slow progress as pro-Assad forces expended considerable effort to pre-empt a major advance on Latakia Province, the government's stronghold.

In control of a large swathe of territory, Jaysh al-Fateh stamped its authority on the northwest and established Idlib as Syria's Islamist heartland. Ahrar al-Sham carefully trod the line between its Salafist, jihadist tendencies and the need to present an acceptable, more moderate face to civilians in its territory, as well as international donors. The group was secure enough for its emir to step down in September, and for its Shura Council to replace him with Abu Yahia al-Hamawi, a senior figure who favoured engagement with the West. Meanwhile, Jabhat al-Nusra kept a low profile, ingratiating itself with other rebel factions and the citizens of Idlib and Aleppo.

Jabhat al-Shamiya, an alliance comprising moderate Islamist factions and secular groups, remained the main rebel force in Aleppo and areas north of the city, although it worked closely with Jaysh al-Fateh. Jabhat al-Shamiya made few advances in 2015, as it was preoccupied with holding off advances by the regime in the provincial capital, and by ISIS near Marea. As part of the Conquest of Aleppo operations room, it announced in June a major offensive against regime-held areas there, but subsequently captured only a few neighbourhoods in the northeast of the city.

However, moderate and secular rebels enjoyed real authority in the south, due to the presence of the Southern Front branch of the Free Syrian Army (FSA). Just as Jaysh al-Fateh was advancing in Idlib, the Southern Front attempted to capture territory in Daraa. Although the moderate group worked loosely with members of Ahrar al-Sham and Jabhat al-Nusra, these rebels were neither as overtly extreme nor as numerous as their northern counterparts. At first, the Southern Front and Islamist groups enjoyed considerable success, capturing Brigade 82 and Bosra from the regime in March, and Kafr Shams in April. But their offensive soon stalled. Despite receiving supplies of arms and ammunition from international backers, and guidance from the US-led Military Operations Center in the Jordanian city of Amman, they failed in two separate attempts to capture Daraa from the regime in June and July. As the provincial capital and most important city in the south, Daraa was heavily garrisoned and the regime was determined to retain some control of the city. As a consequence, Daraa continued to be divided between the rebels and the regime.

In the Damascus suburb of Eastern Ghouta, conservative Islamist faction Jaysh al-Islam remained the strongest force. It continued to resist the government's long-running siege of the area, and made small advances into Adra and Tel Kurdi in September. However, Jaysh al-Islam's future was thrown into doubt in December, when a Russian airstrike killed Zahran Alloush, the group's leader and a dominant figure with de facto authority over all rebels in Eastern Ghouta.

The largest rebel advances of the year were made in northern Syria by an alliance of Kurdish forces and the moderate and secular Arabs of the FSA. The Kurdish People's Protection Units (YPG), the FSA units and the Iraqi Kurdish Peshmerga forced ISIS out of Kobane in January, having almost lost the city in late 2014. Over subsequent months, the YPG and the FSA recaptured much of the territory to the east of the Euphrates river, and at the end of May they made their first advance into Raqqa Province, the ISIS stronghold. Kurdish forces then moved on Tal Abyad, a town of strategic significance for ISIS due to its position on the porous border with Turkey. The YPG captured the town in June, and also took Ayn Issa, further south, ultimately denying ISIS the northwestern part of the border.

In October, Arab fighters established the Syrian Arab Coalition, before allying with the YPG and the Syriac Military Council to form the Syrian Democratic Forces. This large joint force then launched major offensives against ISIS-held territory in both Hasakah and Raqqa provinces, capturing the town of Hawl and Tishrin Dam.

Most of the territory seized by the YPG was subsequently absorbed into the Kurdish-administered territories called Rojava. This move, as well as the perpetration of human-rights abuses by Kurdish security forces against Arabs who had tolerated ISIS, led to tensions between the YPG and some other rebel factions. Accordingly, when the Syrian Democratic Force established a branch in Aleppo Province, Jabhat al-Shamiya perceived this as an expansion of Kurdish influence, leading to brief but violent clashes between the groups. Likewise, fearing the spread of Kurdish influence along its border, Turkey twice fired shells at the YPG, warning the group not to advance further west than the Euphrates. Despite Ankara's concerns, the Kurds were the primary recipients of US support in the year, and all their advances were made with intense air support from the international coalition led by Washington.

In response to the dramatic rise of ISIS in 2014, the US had approved plans for a major train-and-equip programme for rebel forces, on a larger scale than a previous programme that had been run covertly by the CIA. After protracted delays, the programme was launched in May in Jordan and Turkey, but it immediately ran into problems. Firstly, few rebels joined the programme, as many were unwilling to sign a contract demanding that they fight only ISIS and not the regime. Then, in July, the first 54 graduates of the programme in Turkey were deployed into Syria as part of the Free Syrian Army's Division 30, but they were soon attacked by Jabhat al-Nusra and stripped of their US-provided vehicles and weapons. This pattern was repeated in September. The Pentagon acknowledged that the programme had failed the same month, and the effort was suspended in October. It was replaced with a new policy of providing arms and ammunition in bulk to troops fighting under vetted Syrian commanders. Some rebel commanders were extracted from Syria by the US, before being trained and then returned to the country. The first three deliveries of materiel, involving two airdrops in October and one shipment by truck across the Turkish border in December, were received by members of the Syrian Arab Coalition and used to good effect in their fight against ISIS in Raqqa Province.

ISIS displaced

While ISIS steadily lost ground to Kurdish, secular and moderate rebels, the group made significant advances in areas held by the regime. Having lost Kobane and many surrounding villages, ISIS fell back to the Euphrates river, where it was able to hold the line and retain control of key towns such as Jarabulus, despite shelling by the Iraqi Peshmerga. Lacking access to the Turkish border in Raqqa Province after its defeat in Tal Abyad, ISIS was dependent on a supply route running through Manbij and northern Aleppo Province. In an attempt to strengthen this front, ISIS clashed repeatedly with Jabhat al-Shamiya around Marea, making small territorial gains.

In May, the group advanced quickly westwards through Homs Province and drove the regime out of Tadmur, a town on the site of the ancient city of Palmyra (by which name it is often known). There, ISIS established its authority with typical brutality, executing first government soldiers, then those who had worked for the government and, finally, many other people for no apparent reason. As ISIS continued its advance westwards into Jazal oilfield and the Christian-majority town of Qaryatayn, many civilians fled their homes.

The group also made several attempts to capture the districts of Deir ez-Zor still held by the regime. The attacks were driven back until December, when ISIS captured the industrial area in the east of the city. The group was initially more successful in Hasakah in May, when it quickly seized government-held districts in the south of the city. However, YPG forces from elsewhere in Hasakah managed to reverse this advance in June and early July, and to expand their control of the city.

Meanwhile, the US and its allies carried out many airstrikes against the infrastructure of ISIS. Hundreds of air operations targeting oil and gas facilities led to a significant decline in the group's income from fuel sales – from approximately US$90 million per month in late 2014 to US$34m per month in December 2015. The US-led coalition also gathered intelligence using manned and unmanned surveillance aircraft, including the British *Sentinel* and the American *Rivet Joint*, as well as by developing human-intelligence networks. This led to successes such as a May 2015 raid by US special-operations forces targeting an ISIS finance officer in eastern Syria and an attack by a UK *Reaper* aircraft on two British members of ISIS. There was evidence that these and other strikes eroded the group's morale and reduced its operational tempo. But ISIS adapted, and remained capable of concentrating its forces for strategic and tactical offensives in both Syria and Iraq.

The Saudi-led intervention in Yemen led to a reduction in airstrikes against ISIS by the Saudi, Jordanian and UAE air forces. But the anti-Assad forces in Syria received increased support of other kinds from Jordan, Turkey and the Gulf Arab states. These countries made a significant contribution to rebel operations in southern and northern Syria, supplying TOW missiles that were effective against the government's armoured units. Russia's intervention in Syria also caused ISIS to lose territory to the regime.

Propping up Assad: Russia, Iran and Hizbullah

By July, with his forces having suffered several major defeats, Assad was compelled to admit publicly for the first time that the regime was unable to control much of the country. He announced that, due to a lack of military personnel, the government would concentrate on protecting its most defensible front-lines. The regime's growing dependence on Iran and Hizbullah to lead military operations took a toll on both parties, which lost many senior commanders. The rebels killed at least 20 Iranian officers, including Brigadier-General Hossein Hamedani, deputy commander of the Quds Force – Iran's special forces – and eight others who were colonels or more highly ranked. General Qassem Suleimani, head of the Quds Force, was also wounded in a separate incident in Aleppo Province.

With the support of these foreign forces, the regime launched in July a major offensive to seize Zabadani and Madaya in Rif Dimashq, close to the Lebanese border. Zabadani had been the first town to fall to the opposition during the civil war. Jaysh al-Fateh retaliated by besieging Kefraya and Foua, the only two towns in Idlib Province still held by the regime. In all four of the towns, food and medicine became scarce and the incumbent forces ran short of arms and ammunition – but their defences held. In light of the deteriorating conditions faced by civilians there, Iran and Turkey stepped in to mediate between the sides, leading to the agreement in early August of a ceasefire, which was then extended to allow the delivery of food aid. Later that month, the sides implemented another ceasefire. Finally, in late September, a third ceasefire was extended into a six-month truce, under which civilians and injured fighters would be evacuated from the towns. The first evacuations, facilitated by the International Committee of the Red Cross (ICRC), took place in December.

This marked the beginning of a series of localised negotiations between the regime and opposition forces. On 1 December, the sides reached an agreement to end the government's siege of the Waer suburb of Homs, granting the rebels there safe passage out of the area in exchange for overall control. Later that month, the regime negotiated with several jihadist groups, including ISIS and Jabhat al-Nusra, to allow them to withdraw from the southern Damascus suburbs of Qadam, Hajar Aswad and Yarmouk; however, the move was postponed by the death of Alloush, which destabilised nearby Eastern Ghouta. The redistribution of rebel fighters that resulted from these truces, and the regime's entrenchment along key front-lines, helped stabilise the territorial divides of the conflict.

Meanwhile, Assad had relatively few political troubles, although the regime still struggled to recruit young men to fight on its behalf. Conscription caused tension within communities otherwise supportive or tolerant of the government, and thousands of civilians fled to avoid joining the army or paramilitary forces. However, Assad secured the support of leaders of the Druze community, a task made easier by Jabhat al-Nusra's execution of 23 Druze men in Idlib Province in June. This effectively ensured that the rebels would be unable to control Suwayda Province.

Assad received steady backing from Russia throughout the year, in the form of military, economic and diplomatic support – and, in September, the country stepped up its involvement. There was a marked increase in materiel supplied to the regime, which received more warplanes, unmanned aerial vehicles and heavy weapons than ever before. At the same time, the Russian military deployed numerous aircraft and hundreds of military personnel to Syria, taking complete control of Hmeymim naval air base in Latakia Province. Moscow assisted airlift operations by Syrian and Iranian military and civilian aircraft by providing transports and helicopters.

Russia began direct military action in Syria on 30 September, conducting airstrikes against groups it defined as 'terrorists'. However, only about 25% of these operations targeted ISIS, with the vast majority conducted in support of regime offensives. The airstrikes by both fighters and attack helicopters had the desired effect, enabling the regime to recapture significant amounts of territory. This included towns to the south and east of Aleppo city, areas of northern Daraa Province, the town of Qaryatayn in Homs Province, and hills around the border between Latakia and Hama provinces. Russia also deployed military advisers and intelligence personnel to Syria. Although rebel groups did lose ground to the regime, they were surprisingly resistant to their advances.

Moscow's increased involvement appeared to come at the expense of Tehran, whose influence waned in late 2015 – an effect compounded by military exhaustion among Iranian forces. This spurred rumours that, with encouragement from Russia, the regime would disband the National Defence Forces, a militia largely built by Iran, and would integrate its members into the Syrian Arab Army.

International diplomacy and opposition politics

Earlier in the year, Russia had attempted to shape the conflict by mediating between the regime and opposition. The resulting two conferences, held in Moscow in January and April, produced no agreement with the Syrian government despite involving only a small segment of the political opposition. Yet Staffan de Mistura, UN special envoy for Syria, fared no better. After a series of injudicious comments that were perceived by many as sympathetic towards the regime, he lost the trust of the opposition and his plan to establish local truces fell apart.

UN Secretary-General Ban Ki-moon subsequently attempted to restart the Track One dialogue between the opposition and the regime, ordering de Mistura to design a new programme of talks. After consultations with regime and opposition factions, de Mistura's plan for thematic discussions between the sides was approved by the UN Security Council. But the scheme was overtaken by events. The false hope that Moscow's support for Assad was fading created a flurry of diplomacy in August, before the onset of the Russian bombing campaign forced the main international actors to the negotiating table. This led to a conference in Vienna between the US, Russia, Saudi Arabia and Turkey on 23 October, and another – involving 13 other nations, including Iran for the first time – one week later. A degree of consensus between the parties was achieved at the second meeting and, by the time of the third, held in November, Russia had put forward proposals for talks between the regime and opposition. The attendees of the third conference agreed on an 18-month plan, under which talks between the government and opposition would take place in January, and would be followed six months later by a ceasefire and the initiation of an effort to write a new Syrian constitution. An ISIS attack in Paris the day before the conference began, which killed 130 people and caused international outrage, likely speeded the effort to reach a deal.

A shadow fell across the diplomatic process in late November, when Turkey shot down a Russian jet that had strayed into its airspace. The aircraft crashed in northern Latakia Province, where Turkmen rebels killed one parachuting pilot, before also killing the pilot of a Russian rescue helicopter. The incident sparked an angry response from Moscow, which imposed sanctions on Turkey and began intensive airstrikes against the Turkmens. Nonetheless, the new peace plan was endorsed by the UN Security Council in December, and Saudi Arabia began to forge a unified negotiating team for the opposition.

The National Coalition for Syrian Revolutionary and Opposition Forces had elected Khaled Khoja as its president in January. A well-liked and relatively effective leader, he sought to unify the opposition over subsequent months, meeting with a variety of armed factions, including Ahrar al-Sham and other conservative Islamist groups. However, with no end in sight for the conflict, Khoja's government-in-waiting was increasingly marginalised in its role as interlocutor for foreign powers. Western and Gulf states scaled back their funding for the Syrian Interim Government, forcing the organisation to massively reduce the services it aspired to provide, and to lay off hundreds of employees. This shift occurred largely because, as rebel groups within Syria formed effective alliances, their senior commanders became useful contacts for foreign states. In its effort to unite the opposition in the wake of the third Vienna conference, Saudi Arabia held in early December a meeting in Riyadh that brought together 103 representatives of Islamist factions, moderate rebel groups and political parties. They succeeded in forming a semblance of a unified opposition, establishing the High Negotiations Committee and electing Riad Hijab, former prime minister of Syria, as its general coordinator.

But the Democratic Union Party (PYD), the main Kurdish political faction, was excluded from the Riyadh conference. From the beginning of the conflict, the PYD and its military arm, the YPG, had tolerated the presence of regime forces in its heartland of Hasakah, and had often clashed with Arab rebels. Given that this distinct political position alienated it from the factions meeting in Riyadh, the PYD joined with its allies in the Syrian Democratic Forces to form their own political body, the Syrian Democratic Council.

Widespread suffering and a refugee crisis

The humanitarian situation in Syria was dire throughout 2015. The most acute suffering was concentrated in areas under siege – usually from regime forces, but in some cases from the rebels. The UN Office for the Coordination of Humanitarian Affairs estimated that approximately 400,000 people were trapped in besieged territory, to which the UN, the ICRC and non-governmental organisations had little access. This isolation caused many civilians to perish from starvation or a lack of medicine.

Elsewhere, the stabilisation of front-lines allowed for slight improvements in the humanitarian situation. In parts of the northeast that came under the control of fewer, more coordinated groups, humanitarian actors gained greater access to people in need. The UN's 'Whole of Syria' response plan improved the coordination of aid efforts and non-governmental organisations benefited from increasingly well-established networks of local partners, with whom they worked to deliver humanitarian aid and basic services. Local administrative councils, the Civil Defence Force and local community-based organisations were the main providers of services in rebel-held areas. In eastern Syria, ISIS controlled provision of services as part of its ongoing exercise in state-building. In regime-controlled territory, the government continued to provide services, albeit with increasing difficulty.

The Syrian air force continued to launch indiscriminate attacks with barrel bombs and chlorine, causing devastation in heavily populated areas. An ongoing series of air operations targeting busy marketplaces across the country killed many civilians. It also appeared that chemical weapons were used by ISIS for the first time in Syria: the group seemed to have deployed mustard gas in an attack on rebels in Marea and chlorine in strikes on Kurds in Hasakah Province. Tasked by the UN with investigating such incidents, the Organisation for the Prohibition of Chemical Weapons confirmed that chlorine and mustard gas had been used, but did not discuss suspected culprits. As a consequence, the UN Security Council voted in August to pass a resolution establishing a three-person panel of investigators, who would work to identify the perpetrators of all chemical attacks.

The annual death toll of the Syrian conflict declined from 70,000 in 2014 to 55,000 in 2015, but, according to the Syrian Observatory for Human Rights, the proportion of civilians killed remained steady at 24%. However, the Observatory acknowledged the continuing difficulty of obtaining reliable figures for those killed, due to underreporting by armed factions and incomplete reporting by civilian sources, as well as occasional exaggerations. The real number of fatalities was widely believed to be far higher.

The desperate humanitarian situation caused the number of refugees created by the war to rise to 4.6m. In neighbouring countries, the appalling conditions of camps and urban areas dominated by refugees and an almost complete lack of job opportunities encouraged thousands to travel to Europe. Early in the year, many civilians opted to travel through Jordan and Egypt to Libya, where people smugglers provided transport across the Mediterranean in often unsafe boats. However, by late summer, thousands more refugees were travelling to the west coast of Turkey, from where they typically continued their journey to nearby Greek islands in small inflatable craft.

Once they reached Europe, a large proportion of these refugees travelled to Germany, which hosted a growing Syrian community. When German President Angela Merkel announced in August that all Syrians who arrived in Germany could remain, the numbers of Syrians seeking asylum there increased dramatically. Within one week of the declaration, the number of refugees travelling through Lebanon to Europe doubled. Greece and other Southeastern European countries lacked the capacity to process the large numbers of refugees they temporarily hosted, as there had been no advance preparation for the ways in which Merkel's announcement would affect the rest of the continent. Instead, Germany called for other European states to also accept large numbers of refugees, despite the fact that they did not share the country's demographic situation. A compromise was reached to reallocate 120,000 refugees within Europe but, by December, Germany had received 484,000 refugees from Syria, according to the German Interior Ministry.

Nonetheless, the influx of refugees into Europe prompted a rise in the continent's contributions to refugee programmes in the Middle East, with an additional €1 billion pledged to these efforts in September. This increased the funding received by the regional refugee-response plan to 59% of the amount required, but the funding for aid inside Syria remained woefully inadequate, at just 43% of the target amount.

Turkey (PKK)

Key statistics	2014	2015
Conflict intensity:	Low	Medium
Fatalities:	30	2,000
New IDPs:		
New refugees:	2,000	

The dynamics of the conflict between the Turkish government and the Kurdistan Workers' Party (PKK) underwent several shifts over the course of 2015. The sides agreed on a peace deal in the early part of the year, but the agreement was destabilised in the run-up to a national election held in June – and it collapsed the following month, in the wake of a suicide attack on a pro-Kurdish rally. The violence quickly escalated thereafter, with fighting between the security forces and the PKK concentrated in urban areas in southeastern Turkey, particularly the cities of Diyarbakir and Cizre.

Between the June election, which resulted in a hung parliament, and a snap vote held in November, Turkish society became increasingly polarised, as President Recep Tayyip Erdogan and his Justice and Development Party (AKP) intensified their efforts to regain a dominant position. Erdogan was accused of intentionally re-igniting the conflict with the PKK to garner electoral support, while he blamed the rising violence on both the group and the pro-Kurdish People's Democratic Party (HDP).

New hope and a lost peace

Abdüllah Ocalan, the imprisoned PKK leader, called in late February for his organisation to end its struggle. Simultaneously, the HDP and the Turkish government announced a ten-point peace plan. Disarmament was to follow the implementation of the plan. However, early on it became apparent that there was a rift between Erdogan, who publicly opposed the agreement, and Prime Minister Ahmet Davutoglu. The more conciliatory Davutoglu was seen by his critics as too loyal to Erdogan and lacking independence, giving the president significant influence on domestic politics. Erdogan had moved to diffuse the decades-long conflict between the Turkish state and its Kurdish minority during his 2003–14 tenure as prime minister, but as president he appeared reluctant to substantively address the key points of contention between the sides.

Lacking Erdogan's support, the peace process quickly stalled. In May, the PKK cancelled a meeting to discuss disarmament, citing a lack of progress. The government subsequently accused the PKK of using the peace process as an opportunity to rearm, while the group criticised the AKP for constructing military outposts and other facilities in the southeast. Ultimately, both sides showed only limited willingness to work towards peace.

With the AKP having lost its parliamentary majority in the June election, the party was ostensibly required to seek a coalition partner to form a government. Yet its critics alleged that it made no genuine attempt to do so. The HDP had secured 13% of the vote, in the first instance of a pro-Kurdish party gaining representation in parliament under its own banner, rather than through individual candidates elected as independents. A democratic socialist party, the HDP had a relatively broad support base among minority groups with divergent political perspectives.

The AKP leadership was visibly frustrated with the election results. Erdogan was pursuing constitutional reform that would strengthen the presidency, a move that required a two-thirds majority in parliament. The AKP's loss of seats to the HDP therefore derailed his plans. Subsequently, Erdogan's political allies verbally attacked the HDP and he demanded that the immunity of parliamentarians be rescinded, so that members of the party could be prosecuted for their alleged links to the PKK.

Meanwhile, the security situation became increasingly volatile. There was a brief lull in the violence shortly after the peace agreement was announced. However, in early summer, the PKK began to attack construction sites and vehicles in the southeast. Hopes for implementing the deal were dashed on 20 July, when a suicide bomber with reported links to the Islamic State (also known as ISIS or ISIL) killed 32 young activists at a rally held by the Federation of Socialist Youth Associations in the Kurdish-majority town of Suruc. The victims had planned to help rebuild the Syrian city of Kobane, near the Turkish border, and to deliver humanitarian aid to its residents. Kobane had been largely destroyed by fierce fighting between ISIS and a variety of Kurdish groups. While all of Turkey's political factions condemned the bombing in Suruc, the HDP and the PKK accused the government of allowing jihadists to operate in the country, and even of providing logistical support to ISIS that had enabled the attack.

These claims reflected the jihadist organisation's alleged involvement in several other attacks on pro-Kurdish groups in Turkey. Prior to the June election, two bombs had been detonated at an HDP rally in Diyarbakir, killing four people and injuring more than 100 others. Then too, Selahattin Demirtas, co-chair of the HDP, blamed ISIS for the attack, adding that the AKP's policies and rhetoric had created a political climate that fostered such violence. The day after the Suruc bombing, pro-Kurdish protests erupted in Istanbul, Ankara and several cities in the southeast, resulting in clashes between the police and activists. On 22 July, the PKK retaliated for the Suruc bombing by killing two policemen in the border city of Ceylanpinar, claiming that the victims had been working with jihadists in Syria. The Turkish Air Force began bombing PKK and ISIS targets in Syria and Iraq shortly thereafter, focusing primarily on Kurdish forces. The wave of arrests that followed the Suruc bombing involved some supporters of ISIS, but were predominantly directed at alleged members of the PKK and left-wing extremist groups.

By this point, the peace process had been abandoned. Calls by the HDP leadership for a return to the negotiating table were ignored by both the government and the PKK. The latter frequently attacked police and army units on patrol, as well as security installations, and began taking control of some areas of cities in the southeast. The government continued to conduct airstrikes in Iraq and southeastern Turkey, as well as in Syria, to curb Kurdish advances along the border, while coordinating these assaults with ground operations against the PKK.

Urban warfare

One of the most remarkable developments of 2015 was a shift from predominantly rural conflict towards urban warfare. The PKK was once an almost classical guerrilla movement, in the sense that it was mostly based in rural areas, but this changed with its takeover of urban neighbourhoods in southeastern Turkey. There were also reports that several contested neighbourhoods were controlled by the increasingly prominent Patriotic Revolutionary Youth Movement. Having apparently been founded by young PKK supporters in 2013, the movement was often referred to as a PKK youth organisation, although some of its members insisted that they did not receive orders from the older group's leadership.

The PKK fortified the areas under its control by digging ditches and blockading streets. At the same time, the group used improvised explosive devices, mines, car bombs and small arms to attack police and military vehicles and buildings, including police barracks, army outposts and the housing facilities of the security forces. On several occasions, the PKK killed off-duty security personnel.

In reaction to the group's frequent use of improvised explosive devices, the government announced it would provide more mine-resistant, ambush-protected vehicles to the security forces. The Turkish authorities responded to the broader challenge from the PKK by deploying police and army units – including special forces and gunships – to the southeast, and occasionally by relying on village guards, a local militia force. The security forces often blockaded neighbourhoods or whole cities, declaring a curfew before they moved in to destroy the PKK presence in those areas. In late 2015, the focal points of these operations were the Cizre district of Sirnak Province, as well as the Silopi and Sur districts of Diyarbakir Province – all of which saw heavy fighting. There were reports that, at one point in the conflict, tanks fired into the city of Cizre.

The Turkish Air Force also frequently attacked PKK targets in the Qandil Mountains, in northern Iraq, where the PKK has its headquarters. Turkish aircraft also attacked camps and storage bunkers in Turkey. The airstrikes prompted calls from Iraq's Kurdistan Regional Government (KRG) for the PKK to leave the Qandil Mountains, arguing that this was necessary to protect the local civilian population.

Electoral and security dynamics

The violence was closely intertwined with Turkey's electoral process, and the eventual failure of coalition negotiations between the Republican People's Party (CHP) and the AKP paved the way for Erdogan's announcement that elections would be held in November. Meanwhile, party members and government officials used the conflict to attack political opponents.

Societal polarisation became increasingly apparent: AKP supporters twice attacked the Istanbul offices of liberal newspaper *Hürriyet*; and several facilities of the HDP, including the party's headquarters in Ankara, were vandalised and set on fire by anti-Kurdish nationalist protesters. Jihadist attacks also contributed to this polarisation. Kurdish groups were targeted for the third time in the year on 10 October, when two suicide bombers killed 102 people and injured 400 more at a peace rally in Ankara. The event had been organised by trade unions, and was attended by many supporters of the HDP. The government initially stated that several different groups, including the PKK, were suspects in the assault, before conceding that it was most likely carried out by ISIS. One of the attackers was later identified as the brother of the Suruc bomber.

The CHP and the Nationalist Action Party criticised the security forces for their inability to prevent the attack, while the HDP blamed the government directly. The latter subsequently suspended some election-campaign events due to the bombing, as did several other parties. Shortly after the attack, the PKK announced a unilateral ceasefire in light of the upcoming election, a move that had been expected for some time. However, this created only a marginal decline in violence, as the government refused to curtail its security operations. As a consequence, some polling stations in the southeast were moved out of contested areas. The AKP recovered its majority in the November election, although the HDP again crossed the 10% threshold needed to enter parliament. Far from bringing an end to the violence, the election was followed by a resurgence of the conflict that was sustained past the end of the year, despite a renewed effort by Turkish security forces to defeat the insurgents.

Political attacks on the HDP also continued, and criminal investigations were launched against its leaders, including Demirtas and his co-chair, Figen Yüksekdag. Erdogan renewed his effort to rescind the immunity of the party's members of parliament. The HDP reacted to the ongoing security operations with a more definitively pro-Kurdish stance, calling for civil resistance and moves towards regional autonomy, both of which Erdogan defined as acts of treason.

Worsening humanitarian situation

The conflict caused a rising number of civilian causalities, and at least 285 civilians were killed in the year. During the summer, PKK fighters set up checkpoints on highways, shooting people who tried to evade them. The group also targeted civilian infrastructure, and occasionally trains travelling through the southeast – especially in the early stages of the escalation. Later in the year, the centre of activity concentrated on the contested areas of the southeast, where PKK militants defended neighbourhoods they held against government security forces.

During the government-imposed curfews, which grew in length and frequency as the conflict deepened, civilians were prevented from travelling to buy food or seek medical attention. According to the Human Rights Foundation of Turkey, around 1.3 million people were affected by the 52 curfews put in place across seven provinces between August and December. And these measures also brought normal life to a standstill in other ways. Shortly before the security operations in Cizre and Silopi, the authorities advised teachers to leave the area, ostensibly to attend training courses. Civilians fled targeted neighbourhoods before the curfews went into effect, lest they be trapped in their houses. As a consequence, it was reported in December that between 100,000 and 200,000 people had left the southeast since summer. Furthermore, an estimated 300 houses were destroyed in the fighting in Cizre.

The healthcare system was damaged by the movement of many health professionals out of areas affected by the violence, which also regularly prevented ambulances from reaching patients. In December, the health ministry extended the shifts of hospital workers, so that a single shift could last an entire week, and announced that it would deploy armoured ambulances to the areas affected by the conflict.

In line with the concerns expressed by the KRG, civilians in the Qandil Mountains were affected by Turkish airstrikes. Amnesty International stated that there was evidence the operations had led to the deaths of eight civilians and the injury of eight more in the village of Zergele.

The wider conflict also exacerbated the difficulties faced by Turkish and foreign journalists working in Turkey, as evidenced by the attack on *Hürriyet* and the government takeover of Bugün TV prior to the November election. One Dutch journalist was arrested and deported in September, after reporting on the violence in Turkey's southeastern provinces. Two British journalists suffered the same fate, while an Iraqi co-worker of theirs remained in custody at the end of the year, having been charged with terrorism four months earlier.

Amid this hostility and mistrust, there seemed little prospect that Turkey's return to full-blown conflict would be reversed by the peace process. Although the war in Syria, as well as Turkey's strained relationship with Kurdish groups fighting ISIS there, contributed to the political stalemate and escalating violence, the conflict between Ankara and the PKK was primarily shaped by domestic developments such as the suicide attack in Suruc. The elections affected the overall dynamics of the conflict by obstructing Erdogan's effort to establish a stronger presidency, but it appeared that he would persist in working towards this goal. The shift in battlegrounds from rural areas to cities and towns caused significant damage to both civilians and Turkey's infrastructure, yet there seemed to be few signs of compromise on either side. The government continued to conduct assertive military and rhetorical campaigns, while the PKK remained unwilling to lay down its weapons.

Yemen

Key statistics	2014	2015
Conflict intensity:	High	High
Fatalities:	4,000	7,500
New IDPs:	100,000	
New refugees:	350	

After teetering on the verge of collapse for more than a year, Yemen descended into a full-blown civil war in March 2015, leaving the country with two rival governments and two capital cities. The original capital, Sana'a, was under the control of the Houthi movement, a Zaydi Shia clan and militia that seized large swathes of Yemeni territory in 2014–15, provoking a violent power struggle with President Abd Rabbo Mansour Hadi. The Hadi government declared the southern port city of Aden to be the temporary capital in July, after expelling the Houthis from southern Yemen with the assistance of a Saudi-led coalition predominantly composed of Arab countries.

The coalition's victories in the south created new challenges, such as re-establishing Hadi's authority in Aden and other areas that it had ostensibly secured. At the same time, the coalition failed to gain a decisive advantage in its push towards Yemen's central provinces, even after deploying thousands of foreign troops to the country. And attempts to quell the violence through peace negotiations failed to prevent escalation on the ground. This stalemate generated concern that the war would draw out for several years – a dire scenario for civilians facing a severe humanitarian crisis.

The resolution of the conflict was complicated not only by large schisms between the two sides, but also by the presence of a range of smaller actors pursuing separate agendas. This included the organisations comprising the Popular Resistance – southern, Islamist and tribal militias fighting against the Houthis – as well as armed groups such as al-Qaeda in the Arabian Peninsula (AQAP) and a branch of the Islamic State (also known as ISIS or ISIL), which have both thrived amid the chaos.

Domestic power struggle and international intervention

The war in Yemen was often depicted as a proxy conflict between Saudi Arabia and Iran, or a part of a region-wide rift between Sunnis and Shi'ites – and even the parties involved appeared to increasingly rely on these descriptions. Although the war was coloured by regional rivalries and sectarian dynamics, it was in essence a power struggle between the Houthis and the Hadi government. This contest had unfolded as many forces vied to reshape the country in the wake of the Arab Spring, and increasingly relied on force in their efforts to do so. The Houthis used populist rhetoric to tap into long-standing grievances against the government's policies, especially regarding corruption in public spending. But this rhetoric lost some credibility as the Houthis became increasingly belligerent towards civilians.

A significant cause of instability was the process of writing a new constitution, which began in January 2015 and exacerbated tensions between the sides, paving the way for the outbreak of the war. Prior to this, Hadi had entered a fragile power-sharing agreement with the Houthis after they forced their way into Sana'a and actively intervened in daily governance, undermining the little authority held by the leadership. When the parliament tabled a draft constitution that the Houthis deemed contrary to their interests, they responded by detaining Hadi's top aide and besieging the presidential palace. Declaring that Hadi's 'unilateral' actions left him without any claim to legitimacy, the Houthis intensified their insurgency and framed it as a continuation of a popular revolution.

The Houthis' main objection to the draft constitution was its division of Yemen into six federal states, per the recommendations of the National Dialogue Conference held by an array of Yemeni groups, which had concluded in early 2014. The Houthis had previously criticised the division, on the grounds that it would confine them to the landlocked, resource-poor Azal region. The Houthis' incremental advance on Sana'a from their stronghold in northern Yemen was thus an attempt to protect their interests. Despite having fought six wars against the government of Hadi's predecessor, Ali Abdullah Saleh, between 2004 and 2010, the Houthis forged an alliance with him, as he had retained control over key military bases and therefore enabled their territorial expansion.

Hadi resigned from the presidency in January 2015 to protest against acts of violence perpetrated by the Houthis, but reversed his decision a month later after secretly fleeing Sana'a to Aden, his home town. Houthi-aligned militias followed him southwards until they captured Aden in late March, prompting Hadi to escape to Riyadh. By this time, Hadi had become a symbol of Yemen's first post-Arab Spring government and the fundamental challenges it faced.

Capitalising on this image, Hadi called on his allies to drive back the Houthis and reinstate him by conducting a military intervention. Saudi Arabia subsequently assembled a coalition of predominantly Arab countries and launched *Operation Decisive Storm* and later *Operation Restoring Hope*. The intervention was preceded by intense consultations among the Gulf states, aside from Oman, and they provided varying levels of military resources to the effort, with Saudi Arabia and the United Arab Emirates the main contributors. They also carried out an energetic diplomatic campaign to secure the backing of influential countries, and received intelligence and logistical support from the United States, the United Kingdom and France.

Operation Decisive Storm began as an aerial campaign that targeted Houthi forces and supplied weapons to anti-Houthi militias. But as the war continued, the coalition's involvement grew, especially after Saudi Arabia and the UAE adopted an ambitious strategy in late spring to break the stalemate in Aden. In this new phase, called *Operation Golden Arrow*, they co-opted local commanders and launched ambitious train-and-equip programmes for anti-Houthi Yemeni fighters. This rapidly had an effect: they expelled the Houthis from Aden and four other provinces in southern Yemen, allowing the Hadi government to set up in its new capital. The coalition played an active role on the ground, deploying its advisers and special-forces troops, as well as a range of mercenaries. A force from the UAE's elite Presidential Guard, which arrived in Aden in July, was key to supporting the city's recapture and pushing Houthi–Saleh forces back north.

While largely silent about its role, the Emirati force benefited from years of combat experience gained across the region, including in Libya and Afghanistan. Yet the scale and complexity of the mission in Yemen dwarfed previous undertakings, and led to significant losses – as seen in September, when a missile killed around 55 Emirati soldiers, as well as ten Saudi and five Bahraini troops. The composure with which the UAE responded to the incident indicated its growing acceptance of military risk.

Saudi Arabia was believed to have deployed fewer soldiers than the UAE, drawing mainly on its special forces and the National Guard. While the Saudi air force was thought to have maintained relatively high sortie rates, it ran low on munitions, reportedly prompting the UK to divert some of its *Paveway* IV precision-guided bombs to replenish stocks. The US also reportedly replenished Saudi Arabia's supplies in a US$1.29 billion deal. Images published in the international media showed the threat to civilians posed by the use of powerful munitions against targets in urban environments, but also the need for precise targeting information in carrying out such strikes. Military analysts believed that coalition forces were short of tactical air controllers, and as a result US and UK military advisers were assisting the Saudis with targeting.

Western countries faced heavy criticism for their involvement in the coalition, particularly because the air campaign was seen as disproportionately affecting civilians. Although it ostensibly focused on the military assets of Houthi and Saleh loyalists, destroying much of the Yemeni air force and many weapons arsenals, it also struck facilities such as hospitals and food stores, and, according to Human Rights Watch, used cluster munitions. It rebuffed calls for an independent inquiry into these human-rights violations, and Saudi Arabia instead sponsored a UN resolution establishing a national inquiry by the Hadi government. The coalition's success in curtailing an independent inquiry reinforced the perception that the coalition could act with impunity, with tacit support from the international community.

The coalition viewed the advance of the Houthis – and their attempts to unravel the political transition in Yemen backed by the Gulf Cooperation Council – as a *coup d'état* against Hadi and an immediate threat to regional security. Most of the Gulf countries also regarded the takeover as the product of Iranian interference, alleging that the Houthis were financed and supported by Tehran. Although this perspective aligned with Iran's tendency to pursue its interests in the Middle East by using Shia militias as proxies, it ignored the fact that the Houthi movement was supported by some of Yemen's Sunnis.

While there was scant evidence that Iran had provided significant military assistance to the Houthis, the country increasingly lent political support to the movement. Tehran referred to the widespread destruction caused by the war in its condemnation of the intervention, focusing in particular on the humanitarian costs of its campaign, such as the restrictions on aid flows created by the coalition blockade of Yemen. Tensions between the coalition and Tehran rose in May, when the latter refused to allow the Saudi military to inspect one of its ships, which was carrying 2,500 tonnes of humanitarian aid to the Houthi-controlled port of Hodeida. The crisis was defused after Iran agreed to divert the ship to the UN-controlled aid hub in Djibouti, but the incident demonstrated the ingrained mistrust on both sides.

With neither side in the conflict able to gain a decisive military advantage, the United Nations pushed for a political solution and sponsored two rounds of peace negotiations. Yet the talks failed to produce significant results, such as a permanent ceasefire. Instead, they were accompanied by constant violations of the temporary ceasefire agreed on in mid-December, and both sides demonstrated a lack of interest in the UN process.

Foreign diplomats accused Hadi of avoiding the peace negotiations as a means of clinging to power. He only agreed to the talks under pressure from the international community, which called on him to abandon his initial preconditions for negotiations. Most of Hadi's demands related to the implementation of UN Security Council Resolution 2216, which passed in April and called on the Houthis to withdraw unconditionally from the territory they had seized, as well as to relinquish the weapons they had taken from the Yemeni military.

The Houthis eventually accepted the resolution as a basis for negotiations (as demanded by the Hadi government), but they continued to express scepticism about the UN's role. In December, three days after the year's last round of peace negotiations, their leader, Abdelmalek al-Houthi, called on his supporters to continue their 'resistance'. Describing the UN as an instrument of US policy, he dismissed the organisation's efforts to resolve the crisis. Supporters of the Houthi movement argued that the text of Resolution 2216 was one-sided, due to its condemnation of the violent acts of the Houthis but not those of the coalition.

Protracted war among an array of actors

The depth of the coalition's engagement in the war in Yemen suggested that it would struggle to extricate itself from the war. Although the coalition's victories in southern Yemen were initially seen as a turning point in the war, its forces were unable to seize Ta'izz and push beyond Ma'rib, which would have cleared the way for an offensive on Sana'a. Nonetheless, the coalition repeatedly announced its plans to storm the city and dislodge the Houthis. In anticipation of this operation, it set up a base in Ma'rib, directly east of Sana'a, where local tribesmen had cooperated with the coalition.

However, an incursion into the city was likely to be complex due to the challenges of urban warfare, as well as the Houthis' support among some residents and use of weapons typically employed by insurgents, such as sporadic and often unguided rocket fire and improvised explosive devices.

In another indication that the war was becoming protracted, the Houthis opened a front against Saudi Arabia in a border area in northern Yemen. They frequently launched deadly assaults in the southern Saudi provinces of Jizan and Najran, which increased in scale as the war progressed. These cross-border attacks could be seen as an attempt to drag Saudi forces into a direct military confrontation in the area, thereby draining their resources and morale. Saudi forces responded with limited incursions into northern Yemen, but there was no indication of a push into the rest of the country from the north.

Among the various armed groups that were operating in Yemen but were not represented in the peace negotiations, the forces of the Popular Resistance posed a particularly significant obstacle to ending the war. Although their activities largely benefited the Aden government, they were more anti-Houthi than pro-Hadi, and the temporary convergence of interests often masked conflicting long-term agendas. The southern groups of the Popular Resistance provided a prime example of this, as they emerged from a movement that had long advocated independence from northern Yemen. As a consequence, they had refused to recognise Hadi's legitimacy before the Houthi takeover, but subsequently came to view him as the lesser evil.

Although these groups were key to pushing back the Houthis in summer, they framed their activities as a defence of former South Yemen against northern influence and started to hold large-scale independence rallies immediately afterwards. They also denounced the Hadi government's attempts to push southern fighters into the war further north – a move they described as 'cementing the unity of Yemen'. They also dismissed calls to reconstitute the Yemeni army, insisting that an army for South Yemen be established and an international body deployed to create and protect a border between the north and south. Finally, after the Hadi government re-established its base in Aden, the secessionists blocked the road between Aden and Hadhramaut to prevent government officials from entering Abyan.

Unable to adequately control the many groups of the Popular Resistance, the Hadi government and the coalition resorted to a policy of appeasement that had the potential to cause further fragmentation in the so-called pro-Hadi camp. The policy increased the ability of southern groups to pursue demands for secession. Having received shipments of weapons from the coalition, the southern groups were better positioned to take unilateral steps towards independence in the aftermath of the war.

In a further complication, there was evidence that AQAP had participated in some Popular Resistance operations against the Houthis, alongside other Islamist and tribal groups. One example of this came with AQAP's publication of a video documenting urban warfare in Ta'izz, which featured interviews with residents who confirmed the group's activities there. Around 50 of AQAP's fighters controlled key districts in Ta'izz, and its participation in Popular Resistance operations in Abyan may have enabled the recapture of two towns there late in the year. However, there was no ceasefire between AQAP and army forces loyal to Hadi. It was also unclear whether the presence of the group's fighters stemmed from high-level coordination with leaders of the Popular Resistance.

The activities of armed extremist groups, such as AQAP and ISIS, undermined the Hadi government's attempts to stabilise Aden and return to normalcy. Resources were funnelled into reconstructing economic infrastructure in Aden such as the port, the airport and the oil refinery, but attacks by these groups persisted, and the government was unable to gain full control of the city. An ISIS operation against a series of targets associated with the coalition – including Al Qasr hotel, where the government was provisionally based – forced Hadi back into exile in October, although he returned the following month.

Expansion by jihadist groups

The US increased its attacks on AQAP using unmanned aerial vehicles in the first few months of the war and, probably owing to the group's greater visibility, killed its leader, Nasir al-Wuhayshi, along with five other senior figures. Wuhayshi was second only to Ayman al-Zawahiri in the al-Qaeda power structure, but his death was largely a symbolic blow to the group that did little to hinder its activities in Yemen, since its fighters were adaptable and trained to operate independently. Qasim al-Raymi, AQAP's military commander, swiftly replaced Wuhayshi. The group subsequently expanded its area of operations and freedom of manoeuvre, rolling back the Yemeni military's 2014 campaigns against it by exploiting the disorder of war.

Ansar al-Sharia, an affiliate of AQAP, conducted attacks against Houthi positions in central provinces such as Bayda, Ibb and Shabwa. Ansar al-Sharia's success in these operations resulted from its ability to embed itself in existing networks and to establish relations with tribal groups. Seeking to appeal to local actors, AQAP presented the attacks as its attempt to defend civilians against Houthi encroachment. In one instance, AQAP field commander Jalal al-Marqashi did so using sectarian language, claiming that the group was 'protecting' and 'liberating' the Sunni community in the province of Abyan.

AQAP's strategy of winning over the locals became apparent after its fighters deployed to two towns in Abyan in December. The group seemed to have learned from its experiences in 2011 – when it alienated many residents by imposing harsh laws in the province – and adopted a relatively flexible approach, largely staying out of personal affairs and establishing a council to govern in conjunction with local groups. AQAP established franchise groups such as Sons of Abyan as a way to become entrenched in the area without necessarily being at the forefront of political arrangements.

The same strategy was discernible in Mukalla, a port city in the province of Hadhramaut that AQAP controlled from April onwards under Sons of Hadhramaut, in coordination with the Hadhramaut People's Council. After two cyclones battered the city, AQAP's media arm advertised its response plan and released photos of its fighters distributing aid. The group also established religious courts and other administrative bodies in the city, but did not force residents to use them. Although AQAP issued some edicts, including those to ban the drug qat and to segregate men and women at universities, it was unclear whether they were enforced.

However, AQAP's entry into Mukalla prompted some opposition from locals, and demonstrations against its presence peaked in mid-2015. Residents complained about the threat from US aerial attacks on AQAP, and there were reports that the Hadhramaut People's Council pressured the group to withdraw its fighters, to no avail. Some locals expressed their discontent with widespread fuel shortages and power outages, but Mukalla was relatively stable in comparison to other places in Yemen and thus experienced an influx of internally displaced persons.

The power vacuum in Yemen also created opportunities for ISIS, which made new inroads into the country. There was some evidence that ISIS affiliates in Yemen had become stronger than AQAP, attempting to displace the latter from its dominant position among jihadist groups in the country. Established by AQAP defectors who criticised the leadership for being too lenient, the Yemeni branch of ISIS competed for recruits and influence, launching attacks in areas traditionally associated with the older group, such as Hadhramaut. Even ISIS attacks on Houthi positions could be interpreted as a challenge to AQAP. Escalating AQAP's sectarian rhetoric towards the Houthis, ISIS killed hundreds of Shi'ites (both Zaydi and Ismaili) in suicide bombings targeting mosques around the country. Using another well-practised tactic, the group carried out a mass execution and posted a video of the incident online – in this case, the victims were 25 captured Houthis.

While there was no indication of fighting between AQAP and ISIS, both groups sought to delegitimise each other. For example, ISIS-affiliated groups highlighted alleged links between AQAP and Yemeni government officials in an attempt to undermine the AQAP's credibility, calling on its members to join ISIS. Meanwhile, AQAP referred to the 'imminent collapse' of ISIS in Yemen, contending that there was rising disillusionment among its fighters, some of whom were defecting to AQAP. This argument was lent credence in December, when more than 100 members of ISIS, including senior leaders, abandoned the *wali* (governor) of Yemen and asked ISIS leader Abu Bakr al-Baghdadi to appoint a replacement. However, the ISIS Shura Council rebuffed the request, harshly criticising the defection.

Yet these splits notwithstanding, ISIS grew considerably, especially during the second half of the year. This fact was highlighted by the attacks that forced Hadi back into exile, and the group's assassination of the governor of Aden and eight of his escorts in December. As the most significant of a series of assassinations of political and security figures in Aden, the incident increased concern that the city had become a lawless haven for armed groups, whose fighters reportedly controlled key districts and brandished their flags above public buildings.

Internationally, there was little optimism about ending the war in Yemen, which continued to ravage the country. The UN estimated that, between March and the end of the year, the conflict killed 2,800 civilians, including more than 500 children. In June, the UN classified the humanitarian crisis in the country as 'Level 3', reserved for the most severe emergencies, and estimated that 80% of the population faced food shortages. Alongside the blockade imposed by the coalition, the Houthis imposed their own, smaller blockade on the contested city of Ta'izz. The actions of both sides suggested a disregard for the Law of Armed Conflict, as they heavily restricted civilians' access to food, fuel, water and medicine.

Although a long-term ceasefire seemed a distant prospect, questions inevitably arose about Yemen's political future in the aftermath of war, particularly in terms of its continued unity. The largest gains in the conflict were made by armed extremist groups, whose proliferation was a major threat to the security of the country. As Yemen increasingly resembled a failed state and a hub for international terrorism, there was a growing likelihood that the conflict would further destabilise an already volatile Middle East.

Chapter Four

Sub-Saharan Africa

Central African Republic

Key statistics	2014	2015
Conflict intensity:	High	Medium
Fatalities:	6,000	500
New IDPs:	175,000	
New refugees:	165,000	

While still troubled by sporadic sectarian violence, Central African Republic was generally more peaceful in 2015 than in the preceding two years of conflict. The country's leaders made considerable progress in setting a timetable for presidential and parliamentary elections in December, despite the fact that fighting periodically threatened the transitional process. Nonetheless, there remained many unresolved issues between Muslim Séléka rebels and the predominantly Christian anti-balaka militias.

Protests broke out in Bangui in early December, after the Constitutional Court rejected former president François Bozizé's bid to run in the elections. Despite having been sanctioned by the United Nations for his role in the violence, Bozizé remained popular among the country's Christians, and there were reports that he travelled freely in the region, undermining attempts to implement a transition. The elections were also threatened by Nourredine Adam, who had been sanctioned by the UN as leader of former Séléka faction the Patriotic Front for

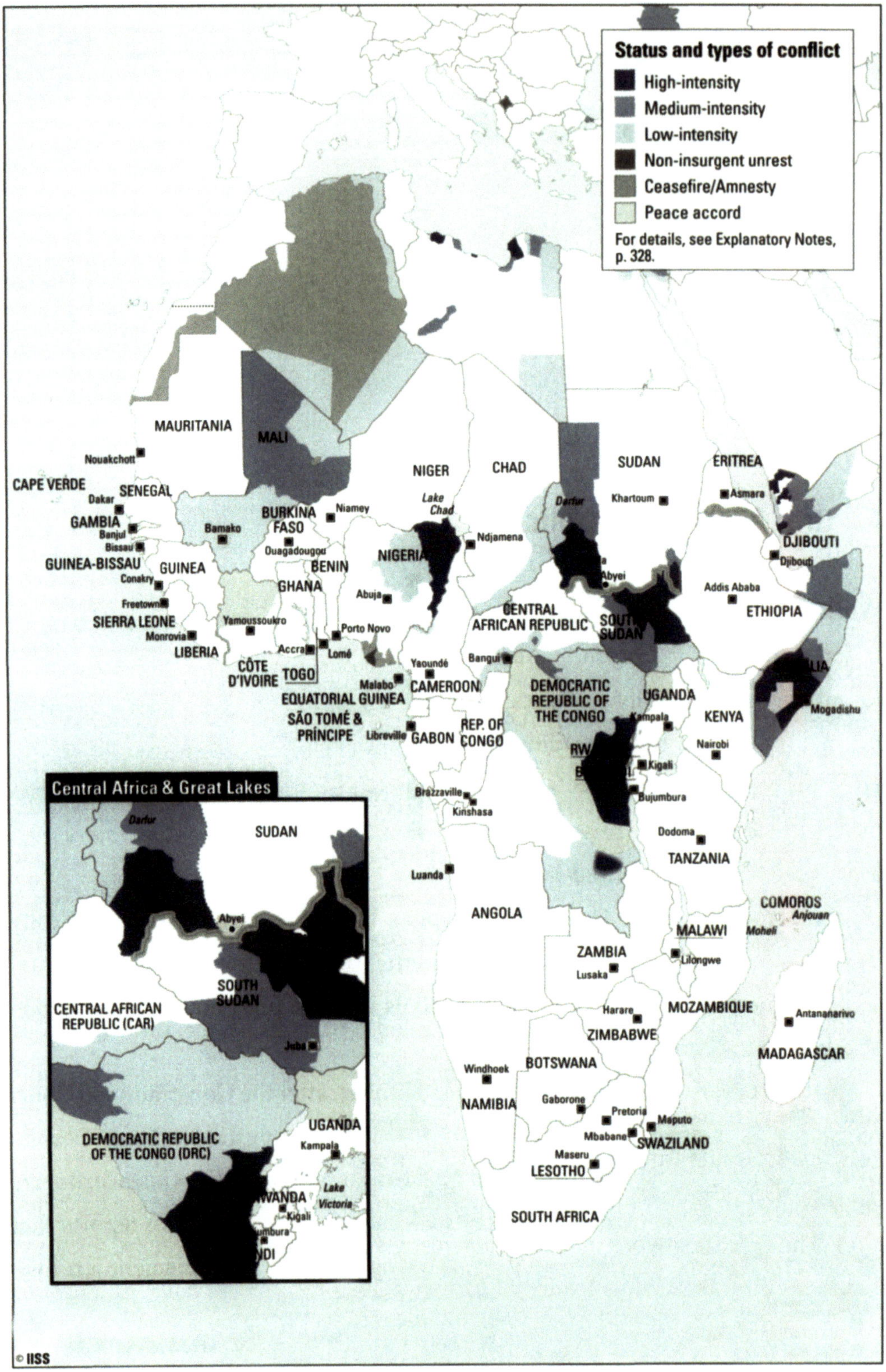
Status and types of conflict
High-intensity
Medium-intensity
Low-intensity
Non-insurgent unrest
Ceasefire/Amnesty
Peace accord
For details, see Explanatory Notes, p. 328.
MAURITANIA
Nouakchott
CAPE VERDE
SENEGAL
Dakar
GAMBIA
Banjul
Bissau
GUINEA-BISSAU
GUINEA
Conakry
Freetown
SIERRA LEONE
Monrovia
LIBERIA
MALI
Bamako
BURKINA FASO
Ouagadougou
Niamey
NIGER
BENIN
GHANA
Yamoussoukro
CÔTE D'IVOIRE
TOGO
Accra
Lomé
Porto Novo
Abuja
NIGERIA
Lake Chad
CHAD
Ndjamena
SUDAN
Khartoum
Darfur
Abyei
ERITREA
Asmara
DJIBOUTI
Djibouti
Addis Ababa
ETHIOPIA
CENTRAL AFRICAN REPUBLIC
Bangui
SOUTH SUDAN
Malabo
Yaoundé
CAMEROON
EQUATORIAL GUINEA
SÃO TOMÉ & PRÍNCIPE
Libreville
GABON
REP. OF CONGO
DEMOCRATIC REPUBLIC OF THE CONGO
UGANDA
Kampala
KENYA
Nairobi
Mogadishu
Kigali
Bujumbura
Brazzaville
Kinshasa
Dodoma
TANZANIA
Luanda
ANGOLA
COMOROS
Anjouan
Moheli
MALAWI
Lilongwe
ZAMBIA
Lusaka
Harare
MOZAMBIQUE
ZIMBABWE
Antananarivo
MADAGASCAR
Windhoek
BOTSWANA
NAMIBIA
Gaborone
Pretoria
Maputo
Mbabane
SWAZILAND
Maseru
LESOTHO
SOUTH AFRICA
Central Africa & Great Lakes
Darfur
SUDAN
Abyei
SOUTH SUDAN
CENTRAL AFRICAN REPUBLIC (CAR)
Juba
UGANDA
Kampala
DEMOCRATIC REPUBLIC OF THE CONGO (DRC)
Lake Victoria
Kigali
© IISS

the Renaissance of Central Africa. He warned that his group would not allow voting to take place in the territory it controlled in the northeast, declaring in late 2015 that the area had become the autonomous Republic of Logone, with Kaga-Bandoro as its capital.

The move followed secret negotiations in Nairobi between former president Michel Djotodia, an anti-balaka group led by Joachim Kokate and Séléka negotiators led by Adam. The talks concluded with the signing in early 2015 of an agreement on disarmament and demobilisation, amnesty for all perpetrators of violence and the removal of Central African Republic's transitional authorities. The deal was denounced by the country's authorities – because they had not been party to the negotiations – as well as the UN and the African Union, all of which called for an end to the Nairobi process that was perceived as unproductive. Regardless of its legitimacy, the agreement was rejected by the Séléka and anti-balaka armed groups on the ground, who continued to carry out acts of violence.

The groundwork for the elections was laid by the Bangui Forum, a summit on national reconciliation and the political future of the country held in May. In a significant step for the transition, the meeting was attended by around 700 leaders from the transitional government, national political parties, the main opposition armed groups, the private sector, civil society and religious groups. At the forum, the armed groups and the transitional government signed an accord on the disarmament, demobilisation, reintegration and repatriation of combatants. The groups also agreed to halt their recruitment of children, and to release all those who had been recruited. Meanwhile, the participants in the broader forum set a timeline for elections and the terms for extending the transitional government's mandate, establishing the structures for justice and reconciliation, which included a national truth and reconciliation commission. However, the optimism created by these successes was tempered by the country's history: since 1980, at least five national debates on peace had been held, all of them followed by periods of crisis.

Protracted violence and flawed peacekeeping

Bangui experienced a sharp increase in violence in late September, following the discovery of the body of a Muslim man who was suspected to have been murdered. The incident led to a cycle of sectarian attacks in which at least 40 people were killed and thousands displaced. Security continued to deteriorate thereafter, and the UN was increasingly vocal in its condemnation of the situation. There were a series of assaults on internally displaced persons (IDPs), five of whom were killed and 5,500 displaced in an attack by Séléka rebels on Batangafo camp in November. Armed groups continued to kill and loot with impunity in the areas they controlled. In light of the growing insecurity, the authorities declared that the elections would not be held as scheduled on 18 October. Although there was a break in the violence in November during a visit to the country by Pope Francis, the fighting resumed shortly after he left, as did the attacks on IDP camps.

Concerns about the persistence of the conflict were compounded by a perceived lack of capacity within the UN Multidimensional Integrated Stabilisation Mission in the Central African Republic (MINUSCA). Several high-level UN officials stated that the mission could only fulfil its peacekeeping role if it received additional personnel and materiel, including troops, police officers and attack helicopters. Although France committed to maintaining 900 *Operation Sangaris* troops in the country until a new president was elected, the European task force in Central African Republic (EUFOR RCA) ended its mission in March. For nearly a year, the latter had provided security at Bangui's airport, as well as in its third and fifth districts, working closely with the African-led International Support Mission to the Central African Republic and then MINUSCA. It was replaced by the European military advisory mission in the Central African Republic (EUMAM RCA), which was tasked with supporting the country's authorities as they prepared to reform the security sector.

Charges of sexual abuse against peacekeepers continued to haunt the international community's efforts in the country. In December, the Paris prosecutor's office interrogated four French soldiers who were suspected of having abused children in Central African Republic between December 2013 and June 2014. According to a report from a three-member panel appointed by UN Secretary-General Ban Ki-moon, the case had been badly mishandled by several senior UN officials, including then-head of MINUSCA Babacar Gaye. Amid several other accusations of sexual misconduct by peacekeepers, Gaye was forced to resign in August.

The scale of the sexual violence occurring in Central African Republic was suggested in November, when the UN announced that it had assisted around 3,800 victims of such abuse, and that 16,900 sexual assaults had been reported between August and October alone. Furthermore, there were 7,517 reported rapes in 2015 (likely a fraction of the actual number).

Civilians were also threatened by widespread displacement. UN Under-Secretary-General for Humanitarian Affairs Stephen O'Brien warned in November that there were at least 465,000 IDPs in Central African Republic, around 65,000 of whom had been displaced by violence in September. Furthermore, around 25,000 Muslims in the PK5 neighbourhood of Bangui remained unable to leave the area without fearing for their lives.

Leila Zerrougui, UN special representative for children and armed conflict, warned in early December that child soldiers were being used as bargaining chips by Séléka groups, which kidnapped them so that they could be released in exchange for political concessions by the government. She added that preventing such crimes against children would be almost impossible so long as the country lacked a functioning judicial system.

In its effort to resolve the conflict in Central African Republic, the international community hoped that the elections would be a turning point for the country, and would bring in a new government able to address its myriad problems. It appeared that much would depend on programmes to disarm, demobilise and reintegrate fighters, and to reform the security sector. However, there was also a desperate need to re-establish the authority of the state, and to unite the warring sides in a national cause. The lack of such authority had allowed the anti-balaka and the Séléka groups to run parallel administrations in the west and east of the country respectively. In this climate, any misstep by a new government had the potential to begin anew the familiar cycles of retaliatory sectarian violence.

Democratic Republic of the Congo

Key statistics	2014	2015
Conflict intensity:	Medium	Medium
Fatalities:	800	1,500
New IDPs:		
New refugees:	44,000	

The conflict in eastern Democratic Republic of the Congo showed little sign of abating in 2015, despite the well-established presence in the country of around 20,000 UN peacekeepers. A number of non-state armed groups continued to operate with impunity, killing many civilians and destroying property in a series of attacks on towns and villages in the east. The violence was often ignored by the country's political establishment, which was preoccupied with elections scheduled for 2016, particularly the issue of whether President Joseph Kabila would succeed in changing the constitution to allow him to run for a third term.

Protracted crisis

The Allied Democratic Force–National Army for the Liberation of Uganda (ADF–NALU) was one of the major perpetrators of violence in the year, and conducted several attacks in the northeastern town of Beni. After ADF–NALU fighters operating in the vicinity of Beni carried out a massacre in April and attempted to down a UN helicopter in May, civil-society organisations called for more effective operations against the group. The massacre was typical in its brutality, and it suggested that the ADF–NALU would continue its activities despite joint operations by the Organization Stabilization Mission in the Democratic Republic of the Congo (MONUSCO) and Congolese forces.

The fight against the group made some progress in July, when its leader, Jamil Mukulu, was captured in Tanzania (he was later extradited to Uganda). Yet this did not prevent the ADF–NALU from wreaking havoc: in late November 2015 alone, it murdered around 30 people in the town of Eringeti, prompting MONUSCO to retaliate using attack helicopters.

The increase in activity involving the ADF–NALU surprised many observers. After defeating the Rwandan March 23 rebel movement in late 2013, MONUSCO and Congolese forces prepared to combat the Hutu militants of the Democratic Forces for the Liberation of Rwanda (FDLR). As a consequence, after the group missed a deadline to disarm, there was much anticipation in early January 2015 that it would come under attack from the joint forces. Having received the backing of the UN Security Council, *Operation Sokola 2* was launched against the FDLR, which was thought to have around 1,500 fighters scattered across North Kivu, South Kivu and northern Katanga. However, MONUSCO did not participate in the operation because the Congolese government insisted that it be led by General Bruno Mandevu, who had been accused of carrying out human-rights violations. The relationship between MONUSCO and the Congolese government had been under strain for some time due to the latter's uneven commitment to defeating the FDLR, and there was concern on the UN Security Council that the disagreement over Mandevu's appointment would lead to a permanent rift. But the incident may have been only a temporary shift in relations with the unpredictable Kabila regime. Forced to operate without international assistance, Congolese troops made only limited progress in *Operation Sokola 2*. The government estimated in August that there were significant numbers of FDLR rebels still operating in North and South Kivu.

Nonetheless, MONUSCO did engage in joint operations with the Congolese military against rebels from the Front for Patriotic Resistance in Ituri (FRPI). In May, around 300 FRPI fighters surrendered to Congolese armed forces and UN peacekeepers in Bukiringi, around 100 kilometres south of Bunia. The following month, MONUSCO deployed ground troops, attack helicopters and unmanned aerial vehicles to support a campaign by the Congolese army against rebels in the northeast. In the joint operations against the FRPI, at least 35 people were killed, while 52 more were wounded and 36 were captured.

As head of the UN force, Martin Kobler reported in late June that these operations had weakened the group. Yet, reflecting on his two years in the role shortly before stepping down, Kobler warned that too many areas of Democratic Republic of the Congo still relied on MONUSCO for security. He was succeeded by Maman Sambo Sidikou, who had led the UN peacekeeping mission in Somalia. MONUSCO's Force Intervention Brigade has special, offensive powers. MONUSCO was mandated to use all means necessary to protect civilians, humanitarian personnel and human rights, as well as to support the government in its stabilisation and peacebuilding efforts.

The insecurity in the east was exacerbated in late 2015 by the activities of armed groups such as the Alliance of Patriots for a Free and Sovereign Congo, Raiya Mutomboki and Mai Mai Kata Katanga. These militias posed a persistent threat to civilians, who were subject to violence, human-rights violations and kidnappings, as well as poverty caused by the theft or destruction of their property. Commenting on the poor humanitarian conditions in the country, Human Rights Watch reported in December that at least 175 people had been kidnapped in the year, the vast majority of them in North Kivu province.

Internal displacement also remained a major problem, with the UN suggesting in August that a total of at least 600,000 people had been displaced and required humanitarian assistance in North Kivu province alone. This prompted Kyung-wha Kang, UN assistant secretary-general for humanitarian affairs, to warn that Democratic Republic of the Congo should not 'fall off the humanitarian radar'.

The country also continued to be plagued by ethnic tension. Following battles between ethnic Lubas and Pygmies in Nyunzu in July, the government asked UN peacekeepers to reinforce troops in the newly established Tanganyika Province (formerly part of Katanga Province). And, in a report on violence in northern Katanga published the following month, Human Rights Watch urged the government to protect civilians. The document focused on one of the worst attacks of the year: an assault carried out in April by Luba fighters, who struck a camp for displaced people near Nyunzu, killing at least 30 Pygmy men, women and children with machetes, axes and bows.

Electoral uncertainty

The country's electoral process was scheduled to begin in October, in preparation for presidential and legislative elections in 2016. However, the provincial legislative elections did not take place as planned, following a decision to reassess the timing of the process handed down by the Constitutional Court in September.

Throughout the year, speculation was rife that President Kabila would violate the constitutionally mandated limit of two presidential terms. With his tenure scheduled to end in November 2016, opposition groups suggested that the changes to the electoral calendar were a delaying tactic designed to clear the way for a third term. Yet Kabila did not comment on his intentions. In an attempt to deal with issues surrounding the elections, he called in November for a national dialogue involving political parties and civil-society groups.

The need to resolve these issues had become particularly apparent in January, when around 3,000 protesters were attacked by unidentified young people at an anti-Kabila rally in Kinshasa. In the violence that followed, 40 people were killed during clashes between protesters and government forces. Shortly after these events, Citizen Front 2016, an organisation composed of 30 representatives from civil-society groups and opposition parties, was created. Its goal was to ensure that no changes were made to the constitution, which had been adopted ten years earlier. A rally was held by Citizen Front in September.

Although it was unclear whether Kabila would run for another term, there was considerable potential for further election-related violence, given the public frustration over the delay. And the situation was made more unstable by the government's reported clampdown on political activities by the opposition. In December, the UN Joint Human Rights Office stated that there had been at least 143 recorded human-rights violations relating to the electoral process, including summary executions and extrajudicial arrests.

Meanwhile, there remained shortfalls in the relationship between the Congolese government and MONUSCO's intervention brigade, to the detriment of civilians terrorised by rebel groups in the east. The most successful phase in the joint counter-insurgency campaign, that which targeted March 23, had not been replicated against other militants threatening peace and stability. The east's many natural resources were still the focus of violent contests involving the insurgents, who were able to exploit them to fund weapons purchases and continuing operations.

Ethiopia

Key statistics	2014	2015
Conflict intensity:	Low	Low
Fatalities:	70	500
New IDPs:		
New refugees:	8,000	

The conflict between Ethiopia's government and the predominantly Muslim, ethnic Somali inhabitants of the country's Ogaden region continued in 2015, though it received little attention from the international media. This lack of coverage was partly due to the ban on foreign media outlets in the region, which has been in place since 2007 and extends to nearly all humanitarian-aid groups. Furthermore, the government has imposed severe restrictions on the activities of non-governmental organisations in Ogaden.

The Ogaden National Liberation Front (ONLF) was founded in 1984. The active phase of the conflict began after the Ogaden Parliament voted for self-determination in the mid-1990s, prompting the Ethiopian government to begin targeting members of the ONLF.

There were unconfirmed reports in early 2015 that the ONLF and the Ethiopian government had resumed a dialogue that had been suspended in 2012, after they failed to compromise on preconditions and a road map for regulation of the conflict. Ethiopian Minister of Defence Siraj Fergessa and his entourage reportedly met in February 2015 with Abdirahman Mahdi and Mohamed Abdi Yasin, the ONLF's head of foreign relations and secretary respectively, with the parties agreeing on most of the principles for guiding the dialogue. It was believed that, unlike previous rounds, the new set of talks would involve the participation of international negotiators. In January, it was reported that Turkey might act as a mediator between the ONLF and the Ethiopian government.

Another promising development was the release in June of two senior ONLF negotiators, who were thought to have been abducted by Ethiopian forces in January 2014. The ONLF stated that their release removed one of the obstacles to further talks.

The security situation in Ogaden deteriorated significantly as a result of Ethiopia's involvement in Somalia, particularly its role in the African Union Mission in Somalia (AMISOM) fighting Islamist group al-Shabaab. In the second half of 2015, there were reports of a considerable increase in clashes in Ogaden. Much of this violence is unrelated to the aspirations of the people in the region, and is instead linked to Somalia.

In September 2015, the ONLF reportedly condemned al-Shabaab for kidnapping and executing two ONLF officers the previous May. The former group suggested that Somali militias were handing over ONLF members and Ogaden refugees to the Ethiopian government. Meanwhile, Somali peace activists accused the ONLF of carrying out several terrorist attacks and assassinations targeting pro-government Somalis in eastern Ethiopia. These incidents allegedly took place in the Aware, Iimay, Qorahay, Danood, Wardheer and Jigjiga districts, among others.

The ONLF attempted to strengthen its influence by merging with the Benishangul People's Liberation Movement, the Gambella People's Liberation Movement, the Oromo Liberation Front and the Sidama National Liberation Front. Announced in October 2015, the move created the People's Alliance for Freedom and Democracy but was unlikely to have much effect on the Ethiopian government's attitude towards the rebels. In a press statement, the new organisation announced that it aimed to help Ethiopians create a 'transitional political order that is based on the consent of all peoples, where the outmoded hegemonic culture of a single group dominating the rest is dismantled and a new just political order is established'.

In the absence of media coverage and activities by non-governmental organisations in Ogaden, it is difficult to evaluate the impact of the conflict on the civilian population. According to reports from human-rights groups such as Human Rights Watch and Amnesty International, civilians are being murdered, raped, imprisoned and tortured by government security forces.

Nigeria (Boko Haram)

Key statistics	2014	2015
Conflict intensity:	High	High
Fatalities:	7,000	11,000
New IDPs:	320,000	
New refugees:	8,000	

On 3 January 2015, Islamist militant group Boko Haram massacred around 2,000 civilians in the northern Nigerian town of Baga. The attack, one of the deadliest in the insurgency's history, displaced tens of thousands of residents of Baga and surrounding areas, most of whom remained unable to return safely at the end of the year. While the Nigerian Army recaptured all the towns that Boko Haram had occupied in 2014, these military victories failed to prevent the insurgents from launching a steady series of assaults. Against this backdrop of persistent violence, the national elections held in March were dominated by the security concerns of voters, the majority of whom denounced incumbent leader Goodluck Jonathan's uneven response to the insurgency. Despite making a belated attempt at decisive action against Boko Haram, Jonathan lost the elections to All People's Congress leader Muhammadu Buhari, ending 16 years of rule by the People's Democratic Party. In a speech delivered at his inauguration two months later,

Buhari pledged to root out Nigeria's endemic corruption, transform the country's economy and defeat Boko Haram.

New government

The appointment of Buhari, a Muslim from the north, transformed the political nature of the conflict. Civilians in the region had grown increasingly discontented with Jonathan, a southern Christian, and saw his vacillation in tackling Boko Haram as one more instance of neglect by Abuja. Given that many southerners also voted for Buhari, it appeared that most Nigerians viewed their government as having failed them in terms of both security and development.

In sharp contrast to Jonathan, Buhari had a reputation as an authoritarian but incorruptible leader, having ruled Nigeria as a military dictator between 31 December 1983 and 27 August 1985 (when he was overthrown in a coup). The peaceful transition of power following the elections therefore generated hope of a new era, with Buhari immediately dismissing the chief of defence, the heads of all three branches of the armed forces and National Security Adviser Sambo Dasuki. Babagana Monguno, a retired major-general with a background in intelligence, was appointed as the new national-security adviser, raising expectations that the counter-insurgency campaign would no longer be hindered by inadequate intelligence gathering. The new government also restored ties with the United States, which deployed an unmanned aerial vehicle in Cameroon to surveil northeastern Nigeria, and sent 300 intelligence advisers to work with the Nigerian military. The United Kingdom also deployed troops to train the Nigerian Army in counter-insurgency, and Abuja accepted offers of assistance from China, Pakistan, Israel, Brazil, France and the United Nations. Meanwhile, Boko Haram abandoned its affiliation with al-Qaeda in March, pledging allegiance to the Islamic State (also known as ISIS or ISIL). Buhari's acceptance of increased support demonstrated that he understood the group could only be defeated through an international effort.

The African Union's Peace and Security Council agreed in January to assist in the fight against Boko Haram by establishing the Multinational Joint Task

Force (MNJTF), which comprised 8,700 troops from Cameroon, Chad, Niger, Benin and Nigeria. The move followed months of resistance from Abuja, which was concerned about the implications for sovereignty of allowing foreign troops to be deployed in its territory. Yet the Nigerian military's severe lack of capacity and the need for regional collaboration had been made clear by a wave of cross-border insurgent attacks that prompted thousands of internally displaced persons (IDPs) to flee to neighbouring Niger and Cameroon. The MNJTF would be mainly deployed to the areas south of Lake Chad, which included the town of Diffa in Niger, as well as Baga and Ngala in Nigeria. Abuja asserted its authority over the force by ensuring that it was led by a Nigerian commander, supported by a Cameroonian deputy and a Chadian chief of staff.

In August, Buhari gave the military a three-month deadline for defeating the insurgents. Although many regarded this as unachievable, particularly given the presence of Boko Haram sleeper cells in Lagos and Abuja, the counter-insurgency campaign made significant progress. By the end of the year, the combined forces had seized all 20 local-government areas once controlled by Boko Haram, and had captured several prominent leaders of the group (following the release of a most-wanted list that featured photos of the insurgents). These successes led the government to state in December that the war had been won, but the claim was hard to reconcile with the violence that still affected civilians in villages and IDP camps across the northeast.

Restructuring the military

Following the massacre in Baga, the newly assembled MNJTF began the major air and ground offensive that led to the recapture of towns controlled by Boko Haram, killing more than 1,500 militants while seizing Gamboru, in Borno State, and four other towns in February alone. By April, the force had destroyed all insurgent camps in the state, and had retaken Bama and Gwoza – the declared capital of Boko Haram's 'caliphate' – as well as Adamawa and Yobe states. Yet despite these successful operations, Boko Haram still conducted deadly attacks in villages in Borno, Yobe and Benue states. The insurgents also stepped up their activities in Chad and Niger, in retaliation for the countries' contribution to the joint task force.

Buhari's restructured military also announced a new strategy for combating Boko Haram, which led to the overhaul of aerial operations and efforts to target the group's supply chain, as well as programmes for collecting and disseminating information. To improve morale, 5,000 soldiers were promoted, and 400 recalled and redeployed. Furthermore, additional compensation was allocated to the children of troops killed in combat.

Nigerian forces arrested 20 prominent Boko Haram leaders between August and December, and the Cameroonian army freed 1,800 kidnap victims in November and December, clearing a series of insurgent camps near their country's border with Nigeria. At least 4,500 civilians held captive by the group were rescued in 2015, although the 276 schoolgirls kidnapped in Chibok in 2014 were not among them. The group continued to carry out abductions as the combined forces advanced.

One of the key challenges in the fight against Boko Haram remained the group's capacity to adapt and change tactics. As the military made large strides in the insurgents' main areas of operation, they resorted to guerrilla warfare to harry and delay its advance. Unable to seize and hold major towns, Boko Haram channelled its brutality into assaults on remote villages – often burning homes, killing indiscriminately and kidnapping young people. There was a sharp increase in its use of suicide bombers, particularly women and children. Of the six suicide bombings carried out by the group in Nigeria and Cameroon in November, five were conducted by women or girls. Similarly, of the nine such assaults conducted in December, six of the attackers were female, while one was a boy aged between ten and 15. The military was able to intercept some of the militants before they detonated the bombs they were carrying, but the majority of suicide bombers struck their targets. It seemed possible that Boko Haram's increased use of women and children indicated weakness within the group, as it resorted to forcibly recruiting civilians chosen at random, as well as hiding bombs near beggars on the street. However, the approach made it more difficult for the military to prevent attacks – as seen in November, when a female bomber posing as a civilian attacked IDPs as they arrived in Maiduguri. The insurgents also exploited weak security in villages near Nigeria's borders with Cameroon, Niger and Chad. Although intelligence gathering improved under the Buhari administration, Boko Haram's fighters remained elusive and unwilling to negotiate, and the military was slow to adapt to their shifting tactics.

Buhari's prompt restructuring of the military emphasised the establishment of a strong force with sufficient arms, foreign support and the capacity to regain trust at the local and international levels. Washington provided aid after Abuja pledged that the military would act with professionalism, lifting a ban on Nigeria's procurement of US arms that had been imposed due to its alleged human-rights abuses. Nonetheless, Buhari largely failed to address the developmental problems that sparked the insurgency, and that continued to exacerbate the violence. The states of the northeast had the highest rates of poverty and illiteracy in Nigeria, problems compounded by widespread disease and severe unemployment. If it were to ignore these factors, the government would likely repeat a mistake made by previous administrations: believing that the group had been conquered, only to watch it rise up stronger and better equipped. The danger of this happening was suggested by the striking imbalance between the government's announcements of military measures and its declarations on short- and long-term developmental goals for the region.

Moreover, it was unclear whether the military had been able to provide security as claimed. According to both the National Emergency Management Agency and Abubakar Aliyu, Yobe's deputy governor and the chairman of the Rehabilitation and Relocation Committee, IDPs were beginning to be returned to their places of origin. Aliyu also stated that the reconstruction of communities would be completed by mid-2016. Yet IDP camps and many of these communities remained insecure – as seen in September, when the insurgents killed ten people at Maikohi IDP camp, in Yola. Similarly, in an attack in January, Boko Haram fighters killed civilians who were attempting to re-enter Baga, which the military claimed to have secured. In December, several IDPs reported that the

National Emergency Management Agency was forcing civilians back to homes in unsafe areas, to demonstrate to local and international actors that the war had been won. Although the deadline for defeating the group likely spurred the military into action, it may have also hindered the implementation of longer-term developmental goals.

Worsening humanitarian crisis

The conflict continued to create a dire humanitarian crisis, having displaced an estimated 3 million people since it began. While IDP camps were set up with the assistance of the government, the UN, non-governmental organisations and local charities, it was reported that only 8–10% of IDPs lived in these facilities, while the majority stayed with friends, relatives or community leaders, or in makeshift settlements they constructed themselves. Given that most government and international aid was funnelled into the camps, the majority of IDPs therefore lacked outside support.

Limited access to primary education remained an important developmental issue in the northeast, and the problem worsened as the conflict intensified. Public schools in Borno only reopened in November, having been closed for more than a year, and locals were reluctant to return to them due to concerns about the safety of their children. Moreover, many schools were severely overcrowded, or else being used as IDP camps. Michael Olukoya, president of the Nigeria Union of Teachers, reported that the conflict had killed at least 600 teachers and displaced 19,000 more. Although Buhari's December budget specified that the government would partner with local and state government to train and recruit 500,000 more teachers, he did not provide further information on plans to tackle Nigeria's education issues, many of which were unique to the northeast (such as militants' use of Almajiri schools as recruitment facilities).

Against this background of limited developmental efforts, the actions of the Nigerian security forces often seemed particularly counterproductive. It was reported in December that, following a clash between members of the Shia Islamic Movement of Nigeria and Nigerian troops in Zaria, soldiers had conducted raids in the area that resulted in the deaths of more than 800 civilians. They were alleged to have committed severe human-rights abuses while hunting those responsible for an attack on a convoy carrying General Tukur Buratai, chief of army staff. This followed Amnesty International's announcement in November that Nigerian soldiers had beaten one young man and injured six others in Yobe. Cameroonian soldiers involved in the counter-insurgency campaign were also implicated in large-scale acts of violence against civilians. Such transgressions had in the past assisted Boko Haram's recruitment of young men, and destroyed any trust between the security services and civilians.

Corruption also continued to threaten civilians, as well as Nigeria's economic, developmental and military goals, with state resources often used to secure political, judicial and private-sector influence. Having fired several senior figures in the defence ministry and the military, Buhari also dismissed the eight executive directors of the Nigerian National Petroleum Corporation and the head of the country's anti-corruption agency. Despite the large defence budget maintained by the Jonathan administration, the military lacked the training and materiel it needed to combat the insurgents. Indeed, at the height of the conflict, there were reports of soldiers fleeing Boko Haram fighters equipped with superior weaponry. Buhari commissioned an investigation into the procurement of arms, leading to the arrest of former national-security adviser Dasuki, who was charged with 19 counts of money laundering and criminal breach of trust. Dasuki had allegedly embezzled more than US$10m during his tenure. In an effort to promote transparency and accountability, the president ordered all government ministries to pay all government revenues into the Treasury Single Account, replacing a complicated accounting system that made it difficult to identify losses of income. Buhari's swift, decisive action appeared to demonstrate the sincerity of his anti-corruption campaign. Yet corruption was deeply embedded in Nigeria's political culture, and in the long term could only be tackled by creating a cultural shift among the country's officials.

Nigeria (Delta Region)

Key statistics	2014	2015
Conflict intensity:	Low	Low
Fatalities:	4	250
New IDPs:		
New refugees:		

The Nigerian economy continued to decline in 2015, due to the sharp fall in the oil price that began in summer the previous year. From a high of US$115 per barrel immediately before the slide, the price had reached US$37 by the end of 2015. The resulting economic slump placed great pressure on Abuja's efforts to develop the impoverished Niger Delta region, as Nigeria's wider economy depended on oil sales for around 80% of government revenue and 90% of exports. The government reiterated its commitment to these efforts, which included vocational and academic training programmes for impoverished youth run by public and private bodies, but progress was slow and mired in corruption. The country's Joint Task Force (JTF) made significant progress in combating illegal oil bunkering and refinement in the Niger Delta, as part of *Operation Pulo Shield*, but these practices remained common.

Initially scheduled for 14 February 2015, Nigeria's presidential elections shaped the conflict in the Niger Delta early in the year. Around 20 people were killed in election-related violence in the region, and former militants there threatened to resume their attacks if the vote forced out the incumbent, Goodluck Jonathan, who came from Bayelsa State. As a consequence, there was widespread fear that victory for his main opponent, All People's Congress (APC) leader Muhammadu Buhari, would jeopardise the Presidential Amnesty Program (PAP). The PAP, signed in 2009 by President Umaru Yar'Adua, granted former militants unconditional amnesty, a monthly stipend of US$326 and access to vocational and academic training programmes in exchange for them handing in their weapons. Having been delayed by the Boko Haram insurgency in the country's northeast, the elections were finally held on 28 March. Jonathan won every state in the Niger Delta but only 45% of the overall vote, falling short of

Buhari's 54%. Following the announcement of the result, some former militants stated that they were willing to work with the new government, while others declared that they could be forced to return to violence, in fear that the new government would end the PAP.

After taking office, Buhari declared that his first objective was to tackle corruption, particularly in the oil industry. As government employees were complicit in the loss of billions of dollars of oil revenue, he quickly dismissed all eight executive directors of the Nigerian National Petroleum Corporation, replacing them with four new leaders, and reduced the number of senior managers at the firm by one-third. The measures formed part of a programme designed to create a lean, transparent and accountable organisational structure, and were followed by the announcement that the president would become minister of petroleum, a newly created position. The anti-corruption drive also led Buhari to dismiss the board of the Niger Delta Development Commission, which was rumoured to have abandoned projects and mismanaged funds.

Violence and development

It became apparent in 2015 that many former militants in the Niger Delta had become reliant on the allowances they received under the PAP, and were not benefiting from the social-development and training initiatives at the core of the programme. This was due to a lack of job opportunities in the region, and the fact that the payments were around triple the national wage, an amount that seemed increasingly unsustainable given the financial pressure on the government. Indeed, in the second half of the year, the allowances of 2,000 former militants were delayed for several months because of a lack of funding.

The resolution of the conflict also depended on the new government's ability to reduce Nigeria's reliance on oil, especially in the Niger Delta, where the extraction of the product was often seen as the only viable source of income. Thus, Buhari emphasised the need for diversification into economic activities that were regarded as less lucrative in the region, primarily agriculture.

However, it remained to be seen whether the new administration would make the long-term investments in other sectors that were needed to effect this change.

The environmental degradation caused by oil extraction and production continued to hinder development in the Niger Delta, prompting Buhari to implement a clean-up programme recommended by the United Nations' 2011 environmental report on Ogoniland. While this was a major step forward, private oil companies delayed the effort. By the end of 2015, Royal Dutch Shell was yet to complete the clean-up operations that followed a spill at the Adibawa oilfield in July. And, according to a report published by Amnesty International in November, the environment was still being damaged extensively by pollution at the sites of four earlier spills identified by the UN, which the company claimed to have cleaned up in 2011. Shell was ordered to pay millions of dollars in fines to various Niger Delta communities in 2015, but there was little evidence of a systematic change in its operational practices.

The safety of civilians was also compromised by low-level violence, particularly kidnappings. Although there was a marked decline in such violence, some former militants often used the threat of force to resolve disputes, leading civil-society organisations to regularly protest against the dangers this created. The recourse to non-violent action signalled that there had been a positive shift in attitudes among residents of the Niger Delta, yet the frequency of the demonstrations indicated the scale of the problems they faced.

The security services achieved some major breakthroughs in their fight against illegal oil bunkering in the region. A ten-day operation by the Nigerian Army culminated on 21 September in the discovery of a large oil-bunkering site in Makoba Beach, close to the Nigerian Ports Authority building in Port Harcourt. The site contained more than 5,000 drums of illegally refined diesel; 150 drums of diesel in 21 boats; 132,000 litres of diesel in four tanker trucks; and a barge holding 165,000 litres of stolen diesel. Soldiers arrested 13 suspects in the operation, and subsequent investigations led to the discovery of information on those who ran such facilities. The site's proximity to the Nigerian Ports Authority suggested that government employees at various levels were involved in illegal oil bunkering. Over the course of the year, more than 350 residents of the Niger Delta were arrested in connection with the practice or attacks on oil infrastructure. Oil bunkering continued to be a profitable industry for a wide variety of actors, including foreign criminal groups, as well as security personnel allegedly bribed to turn a blind eye. It appeared that the Buhari administration could effectively address the problem only by combining security operations with efforts to address corruption, poverty, unemployment and the weak rule of law.

Somalia

Key statistics	2014	2015
Conflict intensity:	High	High
Fatalities:	2,000	4,000
New IDPs:	90,000	
New refugees:	50,000	

The Somali government continued its full-scale war against Islamist militant group al-Shabaab throughout 2015. Although the insurgents had been pushed back considerably since 2011, pockets of Somalia remained under their control. Suicide attacks on government buildings, army bases, hotels and restaurants in the capital, Mogadishu, had become commonplace following the group's adoption of guerrilla tactics in 2011. With the Somali National Army (SNA) fighting alongside the African Union Mission in Somalia (AMISOM), al-Shabaab targeted some of the nearby countries contributing peacekeepers to the international force. Meanwhile, the Somali government was preparing for national elections and pressing on with the establishment of regional administrations. The latter effort was progressing slowly, partly due to delays created by clan rivalries and associated violence.

Setbacks for al-Shabaab

One of the most significant changes in the conflict during 2015 was the apparent emergence of a divide within al-Shabaab, as suggested by a series of news reports on a dispute among the group's leaders over whether to maintain their allegiance to al-Qaeda. Having witnessed the successes of the Islamic State (also known as ISIS or ISIL) in capturing and holding territory in Iraq and Syria, some al-Shabaab fighters had begun to push for an alliance with the newer jihadist group. Yet it was still unclear how this fragmentation would affect government efforts to stabilise Somalia.

Little progress was made in initial efforts by ISIS to gain the loyalty of al-Shabaab, which began as early as March 2015. This was perhaps because, despite having a large number of foreign fighters and diaspora members, the latter group has historically been far less interested in global jihad than in the local issues tied to its conflict with the Somali government. Reflecting this tendency, it took a three-year effort by Ahmed Abdi Godane, al-Shabaab's former leader, to formally ally the group with al-Qaeda.

Nigeria's Boko Haram, an ISIS affiliate, approached al-Shabaab in October 2015, inviting its 'brothers in Somalia' to switch loyalties and join it in carrying out attacks in Ethiopia, Kenya and Tanzania. However, al-Shabaab's leadership rejected the proposal. While sources in the Somali security forces expected that only around 100 of al-Shabaab's fighters would defect to ISIS, they acknowledged the potentially broad appeal of the latter's extensive funds, advanced tactics and international agenda.

Al-Shabaab's leaders also acknowledged this threat, leading some of them to state in November that they would immediately behead any member of the group who was believed to support ISIS. Al-Shabaab's intelligence wing, known as Amniyat, began to pursue, arrest and kill members it regarded as sympathetic to the new group. Despite such threats, there were reports of defections among senior followers of al-Shabaab. The influential Kenya-based preacher Hussein Hassan, once a strong supporter of al-Shabaab, began urging followers to back ISIS. In October, high-ranking al-Shabaab spiritual leader Abdul Qadir Mumin pledged allegiance to ISIS, and was subsequently forced to flee to the Galgala hills of Puntland, in northern Somalia.

Somali Prime Minister Omar Abdirashid Ali Sharmarke announced in June 2015 that al-Shabaab would be defeated militarily by the end of the year. Nonetheless, despite internal discord and ongoing offensives by the SNA and AMISOM, the group remained a potent force, and persisted with its campaign of suicide bombings and attacks on peacekeeping and Somali national army convoys.

Al-Shabaab continued to favour assaults on targets with weak security, such as hotels and restaurants frequented by foreigners, politicians or peacekeepers. A series of attacks on hotels in February, March, April, July and November caused the deaths of more than 100 people, including members of parliament, peace-keepers, politicians, foreigners and other civilians. Some of these operations led to protracted sieges, but in general Somali troops quickly regained control of the affected areas. Al-Shabaab also carried out assaults on well-protected targets, detonating explosives aboard a minibus at the gates of the United Nations' regional headquarters in Garowe, the capital of Puntland. In claiming responsibility for the attack, which killed ten people, the group warned that more UN workers would be targeted unless they left Somalia and 'stopped persecuting Muslims'.

Operation Juba Corridor, designed to remove al-Shabaab from Gedo, Bakool and Bay provinces, continued in July, forcing many militants to regroup in Puntland. Ethiopian and Kenyan forces within AMISOM provided particularly important support to SNA units in the undertaking. Several towns and villages in Gedo, Bakool and Bay provinces were liberated. This advance was seen as a major victory for the Somali government, especially as the towns of Baardheere and Dinsoor had been controlled by al-Shabaab for the preceding seven years. Dinsoor had served as a main base for the group's leadership after Barawe fell to government forces in October 2014. Furthermore, the town could no longer be used as a jihadist recruitment and training centre.

The government also conducted successful offensives in Lower Shabelle Province, Hiraan Province and on Kuday Island, one of the few remaining al-

Shabaab strongholds in the south of the country, as well on a key logistics and operations base near Kismayo, in Lower Juba Province. *Operation Juba Corridor* was supported by AMISOM and US airstrikes on al-Shabaab training bases, and the governor of Bay Province stated in September that these activities had reduced the capacity of al-Shabaab in southern Somalia. This echoed earlier comments by US President Barack Obama, who had argued in April that the battle against the group proved the value of his administration's relatively 'low-investment, light-footprint' approach to counter-terrorism.

Alongside carrying out these air operations, the United States put bounties on the heads of 11 senior figures in al-Shabaab, offering US$250,000 for information leading to the capture of Ahmed Diriye, the group's leader. Intelligence chief Zakariye Ismail Hirsi surrendered to the Somali authorities in May. One of several high-profile al-Shabaab fighters to capitulate in the year, he subsequently urged his former comrades to begin peace talks with the government.

Although it had been significantly weakened, al-Shabaab nonetheless seized control of a number of villages and towns. In early September, the group recaptured positions in Lower Shabelle Province that had been vacated by AMISOM troops. Kurtunwarey district, around 120 kilometres south of Mogadishu, reportedly fell under its control, along with the town of El Saliindi, 65km south of the capital. In October, the group captured two towns in central Somalia, Amara and Ba'adweyn – apparently without resistance, after local militias fled. It also regained control of Buqda, located in an economic hub in Hiraan Province. The group also claimed to have taken over territory in Lower Shabelle Province.

Spillover violence in Kenya

Tensions increased between Kenya and Somalia in 2015, partly because Nairobi held Mogadishu responsible for the rise in al-Shabaab's attacks in its territory, and Kenyan President Uhuru Kenyatta was under growing pressure to counter the group. The devolution of powers to northeastern Kenya, designed to address marginalisation and underdevelopment in the region, was undermined by corruption and clan violence that enabled al-Shabaab to expand and operate there with relative freedom. The militant group was also adept at exploiting some of the weaknesses of the Kenyan security services, which struggled to guard their country's 680km border with Somalia. Nairobi's plan for addressing this vulnerability, which involved building a wall along the border, was one of the main sources of contention in its relationship with Mogadishu.

The threat posed by al-Shabaab in Kenya was significant enough for the US government to issue in July 2015 a travel warning for the country. This came three months after an assault on Garissa University that resulted in the deaths of 142 students, six security personnel and four militants – the most deadly, large-scale and brutal operation by the group since the 1998 bombing of the US Embassy in Nairobi. The international media covered the attack and its aftermath in some detail. National and international media criticised the actions of the counter-terrorism police, as the unit charged with responding to the incident reportedly struggled to access air transport in time.

The Kenyan authorities retaliated against al-Shabaab on many fronts, conducting aerial assaults on the group's positions and freezing the bank accounts of some of its members. The government suspected that al-Shabaab had infiltrated Dadaab refugee camp in northeast Kenya, radicalising and recruiting some of the 335,000 Somalis living there. As a consequence, Kenyan Deputy President William Ruto ordered the UN High Commissioner for Refugees (UNHCR) to close the camp. The UN objected to this request on the grounds that Kenya had obligations to maintain the facility under international law, and that any closure would have significant humanitarian consequences. Somali President Hassan Sheikh Mohamud has repeatedly said there is no link between al-Shabaab and Somali refugees in Kenya.

The Kenya Defence Forces continued to fulfil their peacekeeping commitments in Somalia, despite coming under domestic pressure to leave the country. In September, Kenyan security forces launched *Operation Okoa Boni* to drive out around 300 al-Shabaab militants believed to be hiding in Boni Forest, along the Kenya–Somalia border. The militants were thought to have been responsible for several deadly attacks in parts of coastal and northeastern Kenya.

Progress, obstacles and Vision 2016

The Somali government made considerable progress in implementing Vision 2016, its plan for establishing regional states, revising the constitution and holding national elections. There was some doubt about whether the elections would be held in 2016 as scheduled, even after leaders from regional states, central government and civil-society groups, as well as international representatives, held in December 2015 a National Consultative Forum to determine the electoral process. Nonetheless, the US commended the Somali government for its efforts to reach a consensus on holding an inclusive, transparent and legitimate vote, having reopened its diplomatic mission to Somalia in September – another step in the process of restoring the diplomatic ties severed in its withdrawal from the country in the early 1990s. All participants in the meeting were encouraged by the Somali government to compromise, and to fulfil their commitment to reach a final agreement by 10 January 2016.

These advances occurred despite the fact that the Somali government came to a standstill several times in the year. Nearly 100 members of parliament brought in August a motion of no confidence against President Mohamud, alleging that he had abused his power. This led to two months of political mayhem, brought to a close only after the speaker of parliament suggested that investigations be launched into pursuing the motion, prompting its withdrawal. The instability was created by the complex clan loyalties and feuds interwoven with much of Somali politics, which were particularly evident in the effort to establish regional states.

While significant progress was made in proposing and establishing most of these new administrations, a key component of Somalia's political road map, in many regions the process was plagued by violence. In late 2015, South West State swore in around 90 members of its new parliament – responsible for Bay, Bakool and Lower Shabelle provinces – following a March attack on the administration's presidential building in Baidoa by al-Shabaab. During the assault, the militants had used bombs and small arms to enter the compound (which includes UN offices and an airport), prompting a massive operation in the city by Somali government forces and AMISOM peacekeepers. Al-Shabaab failed to reach the reported target of the attack, Sharif Hassan Sheikh Aden, who as head of the administration was a close ally of the national government.

The interim Jubaland State administration, responsible for Gedo, Lower Juba and Middle Juba provinces, elected in August Sheikh Ahmed Mohamed Islam as its president – a move welcomed by the UN, AMISOM, the Intergovernmental Authority on Development (IGAD) and the European Union. Yet establishing that administration had caused controversy, with a stand-off taking place in May 2015 between the federal government and the recently formed parliament. Mogadishu was unhappy about the criteria used to elect members of parliament, and tabled a motion questioning the legality of the process. The dispute was resolved later that month, after the Somali prime minister visited the region.

There was a considerable deterioration of the government's relationship with moderate Islamist militia Ahlu Sunnah wal Jama'a (ASWJ), which fought against al-Shabaab but opposed the government's attempt to establish a regional administration for Galmudug, as well as the consultations on national elections. While the group maintained that its objections centred on the government's mandate, the proposed voting system and the constitution, they also appeared to relate to its disagreements with Galmudug forces and the Puntland authorities. Despite attempts at ceasefires and peace talks, clashes between Puntland and Galmudug forces in and around the town of Galkayo reportedly displaced 100,000 people between November and December.

The ASWJ also opposed the formation of an administration for Hiraan and Middle Shabelle regions, which the group regarded as under its control. Accordingly, the ASWJ boycotted discussions on the establishment of the administration, which began in early August with widespread support from the government, state ministers, tribal elders, the UN and IGAD.

Deepening humanitarian crisis

The humanitarian crisis in Somalia, one of the largest and most protracted emergencies globally, grew worse in 2015. This was partly due to the fact that, despite tentative political progress and the government's military gains against al-Shabaab, the militant group continued to blockade towns and villages liberated by AMISOM and the SNA. Yet the main drivers of the emergency were low agricultural production partly from poor rains, and the damage to trade caused by conflict and internal displacement. Reflecting this, the UN Food and Agriculture Organization (FAO) said in early September that the number of people in Somalia facing a food emergency had increased from 731,000 to 855,000 in the preceding six months, a rise of 17%. Another 2.3 million people also required humanitarian assistance.

The UNHCR and the EU held on 21 October a conference for pledging aid to those affected by Somalia's conflict, in light of the fact that there were 967,000 Somali refugees in neighbouring states and 1.1m internally displaced persons (IDPs) in the country. At the event, representatives from more than 40 countries and organisations promised €94m (US$105m) in support for Somali refugees in Kenya, with the aim of helping them return home. A pledge was also made to provide training and job opportunities in Somalia for 10,000 returnees.

Adding to the problems faced by civilians in the country, floods in various parts of Somalia had an impact on more than 132,000 people between mid-October and the end of the year, according to the FAO. These included 60,000 civilians in the conflict-affected provinces of Bakool, Bay, Hiraan, Lower Juba, Middle Juba and Middle Shabelle. Meanwhile, international agencies such as Human Rights Watch raised further concerns about the humanitarian situation by accusing AMISOM's troops of committing human-rights abuses against Somali civilians in IDP camps and elsewhere. This forced the mission to state in May 2015 that its forces had no involvement in such violations.

Furthermore, according to the Committee to Protect Journalists, reporters in the country remained under threat, undermining coverage of both politically sensitive stories and humanitarian issues. Although al-Shabaab was the main source of concern, Somali security agencies and other government bodies were believed to be heavy-handed in their dealings with the press, discouraging reports that might show the state in a bad light.

There were increasing reports of villagers and nomadic tribes joining the fight against al-Shabaab. Many of the resulting battles were sparked by the group's attempts to collect 'Islamic taxes' from, and impose sharia law on, these civilians, some of whom lived near the Kenya–Somalia border, and in Gedo and Hiraan provinces. The incidents may have indicated growing resistance to al-Shabaab, both by its own factions and by the population at large.

Although it was difficult to predict how the complex situation in Somalia would unfold, the country's long-term stability seemed to depend on it holding fair, transparent and inclusive elections. Political progress needed to be accompanied by consistently successful military strategies for dealing with an al-Shabaab that threatened to further fragment, as well as economic, social and developmental initiatives for addressing the dire humanitarian conditions that helped the militants gain new recruits.

South Sudan

Key statistics	**2014**	**2015**
Conflict intensity:	High	High
Fatalities:	12,000	3,500
New IDPs:	1,300,000	
New refugees:	500,000	

South Sudan was riven by intense fighting and an accompanying humanitarian catastrophe throughout 2015, although the parties to the conflict made faltering efforts to implement a peace agreement signed in August. Any optimism that surrounded the deal was tempered by the memory of at least seven failed ceasefire agreements, and there remained significant doubt about the willingness of both the South Sudanese government and the opposition to honour its terms. Near the end of the year, the United Nations Office for the Coordination of Humanitarian Affairs reported that 6.1 million people in South Sudan required humanitarian assistance, and that more than 2.4m had been forced to flee their homes.

Troubled negotiations and international pressure

The process leading up to the August agreement between the government and the Sudan People's Liberation Movement–In Opposition (SPLM–IO) was fraught with mistrust and severe violence. Early 2015 saw yet another round of peace negotiations, this time in the Tanzanian city of Arusha, where the government, the SPLM–IO and the Sudan People's Liberation Movement–Former Detainees agreed to reconcile and unify the factions of the SPLM. But at a January summit of the Intergovernmental Authority on Development (IGAD), the leaders of the organisation's member countries (Djibouti, Somalia, Eritrea, Sudan, Ethiopia, Uganda and Kenya) were unable to agree on a power-sharing formula. They had particular difficulty determining the roles in a future unity government of President Salva Kiir – whom the SPLM–IO demanded be removed from office – and rebel leader Riek Machar (formerly Kiir's vice-president). In addition to the intra-SPLM dialogue at Arusha and the IGAD-led negotiations, the Chinese government hosted a consultation meeting in Khartoum on 12 January, bringing together representatives of the government and the armed opposition. The initiative highlighted the growing role played by China in the conflict, as an ally of the government in Juba and an important investor in the South Sudanese economy. Yet the dominant international players in the drawn-out peace process remained the members of IGAD and the troika comprising the United States, the United Kingdom and Norway.

Parliament voted in March to postpone national elections scheduled for June and to extend Kiir's term by three years, following a request by the Kiir administration the previous month. A government spokesman said the extension was aimed at avoiding a power vacuum, should the leadership fail to achieve a peace agreement before the elections. Yet the move was not supported beyond the ruling faction of the SPLM, and by the end of March at least 17 opposition parties had rejected the extension. US Secretary of State John Kerry said in May that South Sudan's future was 'at grave risk of being squandered' because its leaders had failed to reach a deal. At this stage, the process led by IGAD had continued for nearly 18 months, with the organisation consistently unable to cajole or compel either side to stick to its commitments.

At the end of May, Kenyan President Uhuru Kenyatta launched a new initiative to stabilise South Sudan by merging the IGAD-led mediation and the Arusha talks. In early June, considerable pressure from Kenya and South Africa led to five of the ten SPLM former political detainees visiting Juba for several days of talks with President Kiir. The former detainees had been in exile for more than a year and one of them, Pagan Amum, was reinstated as secretary-general of the SPLM on 23 June. Yet despite the optimism generated by the appointment, the composite peace process stalled on 28 June, with the failure of talks between Kiir and Machar in Nairobi. One of the contentious issues during this round of negotiations was the status of the two sides' armed forces.

The negotiations between the government, the SPLM–IO and the former political detainees resumed on 24 July in Addis Ababa. This round of talks took place under a new framework known as 'IGAD Plus', which incorporated representatives from the African Union, the UN, the European Union, Algeria, Chad, China, Nigeria, Norway, Rwanda, South Africa, the UK and the US. A group of the former political detainees was increasingly prominent in the lead-up to the resumed talks, as they sought to bridge gaps between the two main factions of the SPLM. Led by Amum, the group consulted with the sides on the most contentious issues in the discussion, including federalism, security arrangements and the structure of the proposed interim government of national unity. The talks were focused on a draft agreement prepared by mediators in June, which outlined power-sharing mechanisms for central institutions, as well as devolution to states. It also outlined significant governance and economic reforms, including those for the management of oil revenue, and gave the parties 18 months to integrate rebel forces into the national army.

The international community, particularly South Sudan's neighbours, mediated the many talks between the government and the SPLM–IO, and exerted considerable pressure on them to reach an agreement. One example of this came after the 5 March deadline set by IGAD passed and the negotiations again broke down, prompting the UN Security Council to express its 'profound disappointment' with Kiir and Machar. The Security Council then passed a resolution outlining a sanctions regime targeting spoilers, based on a draft resolution that had been put before it on 24 February, but stopped short of imposing travel bans and asset freezes on officials. And the US suggested on 26 June that six rival South Sudanese military commanders become the first people to be blacklisted by the Security Council for threatening peace and stability in the country – a proposal that the UN approved on 2 July.

During a visit to East Africa in late July, US President Barack Obama convened a meeting with leaders in the region, aiming to build consensus on the need to collectively pressure South Sudan's warring parties to finalise a peace deal. Having returned to Washington, Obama warned of further sanctions and other consequences if an agreement was not reached. A few weeks later, the Security Council threatened to impose economic sanctions and an arms embargo on both the government and the SPLM–IO if they did not implement the deal.

Flawed implementation of the peace agreement

Signed in Addis Ababa by Machar on 17 August and Kiir on 26 August, the peace agreement marked the end of one phase of the war. But the widespread brutality against civilians and pronounced ethnic divisions in the conflict meant that South Sudan remained highly unstable, and on 31 August both the government and the rebels issued statements accusing each other of violating the ceasefire. Kiir reportedly described the deal as a 'sell-out' about which he had 'serious reservations', while many observers pointed out that it effectively reinstated the pre-conflict status quo by dividing power among the country's elite and failing to address the political and economic drivers of the war. Low-level conflict linked to cattle raids and reprisal attacks by groups of armed young people continued in many parts of the country, while chronic governance problems and the emergence of a number of smaller armed groups significantly contributed to the instability. Furthermore, the government neither sought justice for the millions of South Sudanese people affected by the conflict nor held to account those responsible for war crimes.

US Ambassador to the UN Samantha Power announced on 4 September that Washington was preparing a new sanctions list in response to the continuing armed clashes in South Sudan. Furthermore, the AU announced on 29 September that it would set up a special court to try people suspected of committing war crimes in the country. This decision came in response to the findings of an AU commission of inquiry formed in 2014 and chaired by former Nigerian president Olusegun Obasanjo, which had concluded that both sides in the conflict were involved in acts of murder, torture, looting and sexual violence. The AU said that the move was designed to find an African solution to the war.

On 8 October, IGAD released a report stating that in the preceding 19 months ceasefire agreements had been violated 29 times by the SPLM–IO and 24 times by government forces. Some of these incidents had occurred after the signing of the peace deal in August. In light of this, IGAD announced on 19 October that former Botswanan president Festus Mogae had been appointed head of a commission set up to monitor the implementation of the peace agreement. The UN reported on 4 November that both the government of South Sudan and the armed opposition continued to stockpile weapons and ammunition, in violation of the peace deal.

The fighting was particularly intense in Unity and Upper Nile states, and there were clashes in Western Bahr al-Ghazal, as well as Western and Central Equatoria. A spokesman for government forces said on 3 October that 52 soldiers and rebels had been killed in Unity State in the preceding few days. The sides accused each other of instigating the armed clashes in Koch and Leer counties, as well as in the area around Rubkotna, in Unity State. The heavy fighting led to an increasingly severe humanitarian emergency, partly because it inhibited the activities of international aid organisations. On 6 October, after being forced to withdraw staff from Leer County by renewed fighting and the looting of its facilities, Médecins Sans Frontières issued a statement calling on the rival factions to respect humanitarian groups. One week later, the Norwegian Refugee Council and the International Rescue Committee said they had suspended their plans to return to central and southern Unity State because of the ongoing violence there. They estimated that 251,000 people in Koch and Leer counties had been cut off from regular humanitarian assistance for nearly six months. While aid organisations were able to return to both areas in December, the violence continued to have a major impact on local communities. An SPLM–IO spokesman said that government forces had attacked the cantonment sites in Upper Nile and Western Equatoria set out for his group in the peace deal – which, if true, constituted a particularly significant threat to the ceasefire. The ongoing violence therefore raised questions about the willingness and ability of South Sudan's leaders to control their fighters, and to enforce the discipline needed to implement the agreement.

The peace deal was also shaken by President Kiir's decision on 24 December to increase the number of states in the country from ten to 28. This created an entirely new set of grievances, since redrawing state boundaries effectively made it impossible for the SPLM–IO to get a share of government positions, as originally agreed. The move was widely criticised by international observers, and a spokesman for the SPLM–IO announced on 26 December that it violated the agreement. Undoubtedly, it radically undermined the agreement by changing the framework upon which the peace deal was constructed.

Thus, it remained to be seen whether Kiir and Machar would fully commit to the agreement. They came under growing pressure to do so from an economic crisis caused by changes in the oil market, which experienced sharp drops in both prices and South Sudan's exports. And this pressure was compounded by foreign powers' increasing willingness to compel them to go along.

Climate of instability

Beyond the primary conflict in South Sudan, low-level violence continued in many parts of the country, apparently as a result of cattle rustling or disputes over land. The instability this created combined with the emergence of various new armed groups – many of which were aligned with neither the government nor the SPLM–IO – to raise further doubts about the prospects for peace. A government official said that, in early October, clashes related to a land dispute between rival communities in Tonj East County, in Warrap State, had led to the deaths of at least 90 people and injury of 144 others. An army spokesman said on 24 October that 30 people had been killed in cattle raids carried out by renegade soldiers in Bor County, in Jonglei State. Lakes State experienced a number of violent incidents, including a battle on 9 November in which five government soldiers were killed and 11 wounded by armed young people. On 1 December, at least 19 people were killed in clashes between the Aliap and Gony communities of Rumbek East County, in Lakes State, while on 5 December armed clashes between factions of the Dinka community in Cueibet County, also in Lakes State, led to the deaths of 14 people and the wounding of at least 35 more. The fighting appeared to have resulted from revenge killings between residents of the administrative divisions of Joth, Akak and Pagor.

Among the new armed actors to emerge in late 2015 were several organisations centred on young people. These included the Arrow Boys, a group based in Western Equatoria and aligned with the SPLM–IO. Meanwhile, in Upper Nile State, a rebel group known as the Tiger Faction New Forces claimed government troops had attacked its base north of Malakal using two helicopter gunships. A new rebel group calling itself the South Sudan Armed Forces claimed on 3 December to have overrun a police post in Idolu, in Eastern Equatoria State, and to have captured three officers. The organisation stated on 8 December that it had seized a military base in Eastern Equatoria, and subsequently announced that more than 125 government soldiers had joined its forces.

These threats to stability were exacerbated by South Sudan's volatile relations with Sudan, with both countries frequently accusing the other of supporting armed opposition groups. Yet the countries made some progress in their negotiations on border demarcation. The AU declared on 20 March that the Sudan–South Sudan Joint Border Commission had agreed to establish an account to cover the costs of demarcation. The commission also approved a final list of cartographers, surveyors and other experts to carry out the task, and adopted terms of reference 'to manage and supervise the boundary demarcation exercise'.

Nonetheless, the sides made little progress in determining the status of the contested Abyei region. Sudanese President Omar al-Bashir announced on 14 March that the area was 'Sudanese and will remain Sudanese', promising to link it to the national electricity grid. Haile Menkerios, head of the UN Office to the AU, subsequently told the Security Council that hardliners on both sides remained committed to a military solution. In early April, officials from Sudan and South Sudan held talks in Ethiopia over insecurity in Abyei, following clashes between the Ngok Dinka and Misseriyah tribes there. But the discussions appeared to have little effect, and shortly thereafter Misseriyah fighters killed at least six Ngok Dinka during an attack on Koldet. Another attempt at negotiations between the tribes was launched in mid-April, with the sides meeting in Aweil, the capital of Northern Bahr al-Ghazal. However, little progress was made in the talks. Due to the ongoing violence in Abyei, and with no resolution of the region's status in sight, the Security Council granted in December a five-month extension to the mission of the UN Interim Security Force for Abyei. The Security Council also called on Sudan and South Sudan to swiftly resume their negotiations on the status of the region.

Overall, there was little prospect of ending the security and humanitarian crises in South Sudan, despite the signing of the August peace agreement. The distance between the government and the SPLM–IO remained significant and, given the levels of mistrust between the sides, the ethnic dimensions of the conflict and the other causes of instability in the country, it was unclear whether they would be able to find new areas for cooperation.

Sudan (Darfur)

Despite faltering efforts to begin peace talks in late 2015, Sudan was troubled by conflict in Darfur, South Kordofan and Blue Nile states throughout the year, leading to a further decline in the country's humanitarian situation. The United

Key statistics	2014	2015
Conflict intensity:	Medium	Medium
Fatalities:	2,500	3,000
New IDPs:	400,000	
New refugees:	1,750	

Nations estimated that 166,000 people in Darfur were displaced in 2015. This was fewer than in the previous year, but the conflict also spread to more areas and became increasingly unpredictable, with much of the violence attributed to criminality and clashes between tribes. In South Kordofan and Blue Nile, ongoing battles between government forces and insurgent group the Sudan People's Liberation Movement–North (SPLM–N) resulted in the displacement of civilians in government-controlled areas, and continued to prevent international aid organisations from reaching territory held by the rebels.

Tentative steps towards peace talks

Having won 94% of the vote in national elections held in April 2015, Sudanese President Omar al-Bashir was able to create a platform for his National Congress Party (NCP) to promote its agenda of 'mature federalism', and to fulfil his promise of achieving peace with Sudan's opposition groups through a national dialogue. Between 2 June, when Bashir began another five-year term, and the start of the dialogue on 10 October, much attention was given to discussions on where the dialogue would be held and whether the armed opposition groups would be allowed to participate in it. The political process was marked by two mutually reinforcing trends: limited moves towards including a broader array of actors in the negotiations, and the increasing factionalisation of the rebels.

Sudan's national-dialogue body, a composite of opposition parties and pro-government parties known as the 7+7, announced in mid-July that it was ready to meet with the rebel groups, having previously refused to do so in Addis Ababa in April 2014 under the auspices of the African Union High Level Implementation Panel. The national dialogue reflected Sudan's overall power dynamics, in the sense that the armed opposition was confined to both the country's geographical periphery, Darfur and the Three Areas (Abyei, South Kordofan and Blue Nile states), and its political margins. The shifting loyalties and internecine squabbles that had long roiled Darfur's rebel groups were evident in early August, when two elements within the Dabajo faction of the Justice and Equality Movement (JEM) – which had split from the original organisation in 2012 – stated that they would form separate groups. However, moves to include the armed opposition in the dialogue were strengthened the same month with Bashir's statement that he would provide guarantees against arrest to the Sudan Revolutionary Front (SRF), an umbrella organisation for rebel movements from Darfur, Blue Nile and South Kordofan. The SRF had previously said that it would not participate in discussions inside Sudan, partly due to the death sentences handed down *in absentia* in 2014 to the leaders of the SPLM–N, Malik Agar and Yasir Arman.

The national-dialogue body announced in September that the government had agreed to release all political detainees, end its censorship of the press and ensure the freedom of all parties during the dialogue process. The announcement came following the arrest in August of approximately 17 members of the opposition, most of whom were affiliated with the Sudanese Congress Party. President Bashir also proposed a limited ceasefire agreement between government forces and the SRF. Despite having boycotted the dialogue process, the main SRF factions – including the SPLM–N, the JEM and the faction of the Sudan Liberation Movement led by Minni Minnawi (SLM–MM) – announced in September that they were willing to sign a six-month agreement on the cessation of hostilities with the government (they went on to do so on 18 October). A few days after the announcement, President Bashir said that he would pardon any rebel leaders who participated in the dialogue process, and he declared a two-month unilateral ceasefire in the conflicts in Blue Nile, Southern Kordofan and Darfur.

Nonetheless, clashes continued throughout the national-dialogue conference, which was boycotted by many of Sudan's major political and militant opposition groups. Moreover, in late November, a Khartoum court handed down death sentences to 18 members of the SLM–MM, who had been captured during fighting in North Darfur in 2014. A few weeks later, separate talks mediated by the African Union began in Addis Ababa, allowing the Sudanese government and some of the armed movements to discuss a proposed cessation of hostilities in Darfur, South Kordofan and Blue Nile. Unsurprisingly, they were unsuccessful, and the African Union suspended the process after the parties failed to reach any agreement.

Sudan's troubled relations with the International Criminal Court (ICC) deepened in November, when the court stated that it would refer the country to the UN Security Council for failing to arrest Darfuri rebel leader Abdallah Banda Abakaer Nourain, whom it accused of war crimes. This followed the ICC's issuance in September 2014 of a warrant for his arrest, citing his alleged involvement in an attack on African Union peacekeepers in North Darfur seven years prior.

Ongoing military clashes

Reports on the battles between armed opposition groups and government forces in Darfur, South Kordofan, Blue Nile and Sennar were often contradictory, and in many instances there was no independent confirmation of casualty numbers, the outcome of the fighting or even that the incident had taken place. International human-rights and aid organisations reported that government offensives included aerial bombardment that killed or displaced many civilians.

During a series of clashes in Darfur's Jebel Marra region in January, 22 people were killed and many more were wounded. A large government offensive against forces from the faction of the SLM led by Abdel Wahid al-Nur included airstrikes that destroyed several villages in Golo, in Central Darfur, and a ground offensive led by the paramilitary Rapid Support Forces (RSF). Meanwhile, in South Kordofan and Blue Nile, there was intensifying conflict between the Sudan Armed Forces (SAF) and the SPLM–N. The latter claimed in mid-February to have taken the towns of Jineziya, Angartu and Um Turuq-Turuq, while battles around the town of Kalogi, in South Kordofan, in early March caused dozens of casualties and left hundreds of people homeless. The UN reported that more than 20,000 people were displaced due to fighting in South Kordofan in March. On 22 March, the SPLM–N carried out an assault on

Al-Gabanit garrison, in Blue Nile, leading to a battle between the rebels and the SAF.

The fighting grew worse in April, as rebel groups attempted to disrupt the presidential elections by stepping up their attacks. Early in the month, SPLM–N fighters reportedly engaged in several battles with government forces in South Kordofan. A spokesperson for the SPLM–N stated that the group had carried out an attack on the town of Farishaiya on 29 March, killing 28 soldiers and seizing a number of weapons. The JEM was also involved in clashes in South Kordofan, and was believed to have been behind a 21 April attack on a SAF garrison. The JEM stated that it had killed 31 soldiers and destroyed eight military vehicles in the assault. Following the election, the government carried out offensives against the SPLM–N with renewed vigour, eager to make gains against the group before the onset of the rainy season. In response to Darfuri rebel groups' intervention in the conflict in South Kordofan and Blue Nile, forces from the RSF also deployed in early May to the former state. Defence Minister Abdul-Rahim Mohamed Hussein claimed that together the SAF and the RSF would 'clear out the rebels'.

In April, the Sudanese military accused South Sudan of supporting Darfuri rebels, contending that the JEM was using the Bahr al-Ghazal region as a base for its operations in Sudan. Although Khartoum had made similar claims in the past – primarily in connection with the SPLM–N, which was once a faction of the party that went on to rule South Sudan after it became an independent state – the accusation underlined the fact that the conflicts in the two countries appeared to be merging. Like the 2014 reports that the JEM had fought alongside South Sudan's forces in Unity State during that country's civil war, Khartoum's claim highlighted some of the cross-border dynamics of the conflicts near Sudan's southern border.

Following a lull during the rainy season, the fighting intensified in August before diminishing again in subsequent months, after a ceasefire was announced by President Bashir. The SPLM–N claimed to have seized control of Gadier and Abu Grain garrisons in South Kordofan on 18 August. The following day, the rebels said they had repulsed an attack by government forces who were attempting to capture the strategically important area of Jebel Kolgo, which lies 30 kilometres southwest of Ad-Damazin, Blue Nile's capital. Despite the 18 October declaration by the SRF factions that hostilities would cease for six months, the group continued to engage in sporadic clashes with Sudanese forces. One such incident occurred in late November, with a spokesman for the SPLM–N claiming that its bases in Blue Nile had come under attack from government troops.

The UN reported on 8 April that 14 civilians had been killed and 18 wounded in an assault conducted by Sudanese warplanes in West Darfur. In September, the SPLM–N stated that government forces had carried out an aerial bombardment of several villages in Blue Nile, killing one child and wounding several others. Later that month, the rebel group reported that the government's aircraft had bombed several areas in the Um Dorain region of South Kordofan, and that its artillery had struck targets near Kadugli, the state capital, despite the ceasefire.

The Security Council granted in late June a 12-month extension to the mandate of the joint UN–African Union Mission in Darfur (UNAMID). The Sudanese government continued to insist that the mission was no longer necessary, and declared that it would pursue negotiations on its departure from Sudan. Yet UNAMID peacekeepers were periodically targeted by attacks throughout 2015. In April, approximately 40 gunmen riding horses and camels opened fire on Nigerian UNAMID troops protecting a water supply at Kass, in South Darfur. The following day, another UNAMID patrol came under assault, resulting in the injury of one peacekeeper. UNAMID troops also reported in July that they had repelled an attack by unidentified gunmen, and in September a South African UNAMID soldier was shot dead near Fashir, North Darfur's capital.

Worsening humanitarian situation

An overriding feature of the conflicts in Sudan has been the deteriorating humanitarian situation, and the inability of international aid organisations to reach many of the civilian communities most affected by the violence. The UN

Office for the Coordination of Humanitarian Affairs (OCHA) said in August that, despite years of international assistance, Sudan's 'humanitarian needs remain acute and in some cases are worsening'.

The escalation in early January of operations by government forces against rebel groups in and around Jebel Mara led to a significant displacement of people in North Darfur and South Darfur. The UN reported in June that at least 78,000 people and possibly as many as 130,000 had been driven from their homes during the Sudanese army's offensive in Darfur in 2015. Violence in Blue Nile also continued to displace civilians. The OCHA reported in mid-April that approximately 20,000 people had been displaced in Geissan, following an escalation of fighting between the SPLM–N and government forces in the area. The conflict in South Kordofan had similar humanitarian repercussions: in April, UNICEF condemned in 'the strongest possible terms the continued loss of innocent lives in Sudan as a result of armed conflict'. In the two weeks leading up to the announcement, at least ten children between the ages of two and 15 had been killed in the state.

The intensity of the fighting early in the year led Médecins Sans Frontières to pull out of parts of Blue Nile, forced closure of its facilities in East Darfur and created administrative obstacles that halted its work in South Darfur. The OCHA estimated in late August that, overall, the conflict had led to the displacement of 3.1 million people, 2.5m of whom were in Darfur and 538,000 in South Kordofan, West Kordofan and Blue Nile. The UN also pointed out that instability near Sudan's borders is adding to the humanitarian burden, with thousands of civilians seeking asylum and refuge in the country. Many of these people are fleeing the war in South Sudan, and are settling in refugee camps in Blue Nile.

Several international organisations reported serious human-rights violations in Sudan in 2015. Human Rights Watch stated in mid-April that it had evidence of Sudanese forces' use of cluster bombs in civilian areas in South Kordofan's Nuba Mountains in February and March. This was later corroborated by the Geneva-based Landmine and Cluster Munitions Monitor, which reported in September that Sudan's armed forces had used aerial cluster bombs in Darfur and South Kordofan during the first half of 2015. Human Rights Watch declared in March that the government had carried out indiscriminate air attacks in the area, which had killed and maimed children. At the beginning of August, a report by Amnesty International found that aerial bombardment and the deliberate targeting of schools and hospitals by Sudan's armed forces constituted war crimes. It also said the government had refused to allow humanitarian aid into areas controlled by the SPLM–N, exacerbating the crisis. However, Sudanese Foreign Minister Ibrahim Ghandour stated in mid-October that the government would allow international aid groups access to South Kordofan and Blue Nile, provided that they did not provide assistance to the armed opposition groups fighting in those areas.

Reports by Human Rights Watch also documented a continuing crack-down on the political opposition by Sudan's National Intelligence and Security Service, which involved the arrest of dozens of people. Another report by Human Rights Watch maintained that the RSF militia had carried out mass rapes and killings during its two counter-insurgency campaigns in Darfur. The UN High Commissioner for Human Rights also detailed crimes against civilians in Darfur, declaring that the Sudanese security forces had abducted and killed civilians with impunity.

While the humanitarian crisis in Sudan has been caused primarily by armed conflict, environmental factors have also driven displacement and worsened food security. The UN Food and Agriculture Organization reported in late September that there was a high risk of crop failure and poor harvests in Darfur, South Kordofan and West Kordofan in 2016, due to low rainfall. The OCHA also reported that civilians in seven out of Sudan's 18 states were experiencing global acute malnutrition rates at or above the emergency threshold of 15% (three of these states were largely unaffected by the conflicts).

The conflicts in Sudan alternately flared and receded in 2015, and the year ended with another failed attempt to reach a ceasefire agreement between the government and some of the rebel groups. Despite fluctuations in the intensity of the fighting, the underlying drivers of the conflict and the key disagreements between the government and the armed opposition groups remained the same. If anything, the sides grew further apart due to the marginalisation and collective punishment of civilian populations, mainly by government forces and pro-regime militias, as well as the impact of drought and the inability of international aid organisations to reach communities affected by the violence.

Chapter Five

South Asia

Afghanistan

Key statistics	2014	2015
Conflict intensity:	High	High
Fatalities:	7,500	15,000
New IDPs:	155,000	
New refugees:	24,000	

The conflict in Afghanistan escalated sharply in 2015, as the Afghan National Security Forces (ANSF) assumed full responsibility for the country's security following NATO's drawdown the preceding year. Emboldened by the departure of most Western troops, the Taliban conducted an aggressive countrywide insurgency campaign that led to a series of major victories, including the brief but symbolically important capture of Kunduz in September. A nascent peace process was suspended on 30 July, when the group announced the death of its leader, Mullah Omar, and his successor, Mullah Mohammad Akhtar Mansour, was forced to focus on asserting control of the insurgency. According to an unpublished UN assessment, by October, the Taliban had established a presence in more parts of Afghanistan than at any time since 2001, and the security threat in half of all districts was either 'high' or 'severe'. At the same time, militants who had pledged allegiance to the Islamic State (also known as ISIS or ISIL) emerged as a new threat in the east of the country. Faced with unprecedented civilian casualties

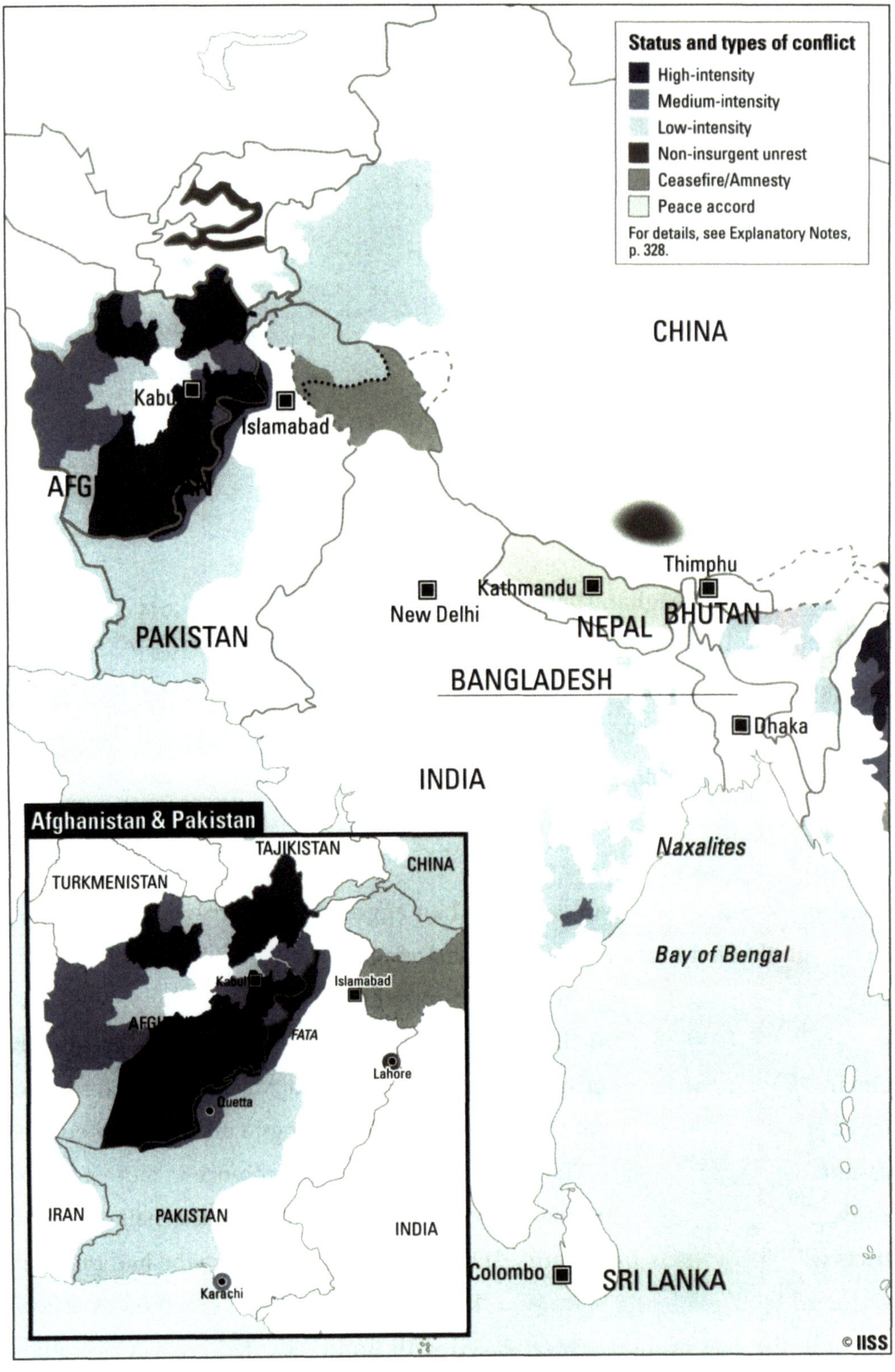
Status and types of conflict
High-intensity
Medium-intensity
Low-intensity
Non-insurgent unrest
Ceasefire/Amnesty
Peace accord
For details, see Explanatory Notes, p. 328.
CHINA
Kabul
Islamabad
PAKISTAN
Thimphu
Kathmandu
New Delhi
NEPAL
BHUTAN
BANGLADESH
Dhaka
INDIA
Naxalites
Bay of Bengal
Colombo
SRI LANKA
Afghanistan & Pakistan
TAJIKISTAN
TURKMENISTAN
CHINA
Kabul
Islamabad
FATA
Lahore
Quetta
IRAN
PAKISTAN
INDIA
Karachi
© IISS

and around 15,000 fatalities overall – double that in 2014 – the United States and its NATO allies delayed the withdrawal of their remaining forces.

Weak security institutions

On 1 January, NATO formally replaced the International Security Assistance Force (ISAF) with *Operation Resolute Support*, a mission to train, advise and assist the ANSF that was initially expected to last until 2016. The US counter-terrorism effort *Operation Enduring Freedom*, separate from ISAF, was replaced the same day by *Operation Freedom's Sentinel*. As a result, more than 12,500 foreign troops remained in Afghanistan. These included 9,800 US soldiers, 6,800 of whom were attached to the NATO-led coalition, while the remainder were tasked with counter-terrorism operations. Although *Resolute Support* was not classified as a combat mission, international forces continued to provide combat support such as airstrikes, and were authorised to use force when coalition troops were under threat.

However, the transfer of responsibility to the ANSF led to growing international concern about deteriorating security in Afghanistan, epitomised by the Taliban's capture of Kunduz. US President Barack Obama declared in October that he would halt the withdrawal of American soldiers from the country until late 2016, with the aim of reducing the size of the force to 5,500 troops by early 2017. Shortly thereafter, the other members of the coalition announced similar delays to the withdrawal of their forces.

According to the Pentagon, a record 7,200 members of the ANSF were killed between January and mid-November 2015. Due to these heavy losses, and to the desertion of many other troops, the ANSF struggled to meet its recruitment targets. The US special inspector general for Afghanistan reconstruction reported that, between May and October, the total strength of the ANSF fell by more than 9,300 personnel, to around 323,000 (including police officers and civilians working for the Ministry of Defence). This left the Afghan National Army and the police with 170,000 and 146,000 personnel respectively. The Ministry of Interior Affairs stated in September that it would investigate the actual number of personnel commanded by the ANSF, due to suspicions that many people on its payroll did not exist – significantly affecting performance on the battlefield.

Partly as a response to this, Kabul resorted to using militias to combat the insurgency in the north. The participation of these groups made the battlefield more complex, increasing criminality and damaging popular support for pro-government forces. The reliance on militias also meant that local power struggles were more likely to delay progress on the battlefield. It was reported in October that Kabul planned to expand the 30,000-strong Afghan Local Police militia programme, adding another 15,000–30,000 troops.

Private militias also endangered state stability in other ways. Both warlords and former jihadist commanders in the anti-Soviet resistance continually criticised Kabul for excluding them from politically influential positions, as well as for displaying indecisiveness on security issues, particularly an attempted rapprochement with Pakistan. Vice-President Abdul Rashid Dostum, an ethnic Uzbek and former Northern Alliance warlord, reportedly attempted in January to form a 20,000-strong anti-Taliban force, consisting of northern fighters (non-Pashtuns) loyal to him. In August, Dostum personally led ANSF operations against the Taliban in Faryab Province.

Meanwhile, President Ashraf Ghani continued to face difficulties in forming a government, partly due to his rivalry with Chief Executive Officer Abdullah Abdullah. Ghani had been inaugurated in September 2014, but cabinet nominations were only finalised the following March, with Masoom Stanekzai appointed defence minister in a temporary capacity in May. A parliamentary election scheduled for June 2015 was postponed until the next year, after Ghani and Abdullah were unable to agree on reforms to electoral bodies. These reforms were particularly important because the United Nations viewed them as a prerequisite for providing financial and technical aid to electoral institutions.

Taliban advance

The Ministry of Defence said in February that it would launch 'clearing operations' across Afghanistan to pre-empt the Taliban's traditional spring offensive.

The most significant of these was the 50-day *Operation Zulfiqar*, the ANSF's first independent special operation, which aimed to remove the insurgents from the Sangin district, in southern Helmand Province. The government claimed that the undertaking was a success, but a Helmand legislator contradicted this view, while other provincial officials and civil-society organisations criticised its impact on the local population. And, soon after the army departed Sangin, local police reportedly began deserting their posts due to a lack of support. Problems with planning and coordination among security agencies persisted throughout the year.

The Taliban claimed to have carried out 327 attacks across Afghanistan on 24 April, the first day of the spring offensive, though there was no way to confirm this. The fighting season was the worst for many years, judging by civilian and ANSF casualties and the countrywide spread of insecurity; a US official stated that, between January and May, the ANSF suffered 70% more fatalities than in the same period the previous year, when NATO forces still had a significant combat role. The security forces' lack of significant airpower was partly to blame, and as a consequence they struggled to counter the Taliban's large-scale offensives. The insurgents intensified their attempts to seize territory, and their campaign of suicide attacks in major cities. Minister of the Interior Noor ul-Haq Ulumi said on 13 June that 12 of Afghanistan's 34 provinces faced 'high' security threats, while nine faced 'average' threats.

The Taliban put northern Afghanistan at the heart of its strategy in 2015, conducting nearly as many operations there as in its southern heartland. On 25 July, the group captured a strategically important police station at Tirgaran, in Badakhshan Province, seizing food, weapons and ammunition. Around 110 police officers stationed there were forced to surrender to the insurgents after days of fighting without reinforcements (local elders later negotiated their release). Kabul subsequently launched a probe into the fall of the base, which sat on an important supply route from Pakistan. Badakhshan officials announced on 20 August that the Taliban had blocked these routes for several communities in Ishkashim, Shighnan, Wakhan and Zebak districts, threatening their inhabitants with food shortages and possibly starvation. Despite having had little influence in Badakhshan in past years, the insurgents contested large parts of the province in 2015. This advance was likely aided by the Taliban's adaptation to local ethnic and political dynamics, as shown by its use of non-Pashtun fighters.

The Taliban's brief conquest of Kunduz, Afghanistan's fifth-largest city, demonstrated not only the insurgents' advanced tactics but also the ANSF's lack of coordination and morale. Kunduz became the first provincial capital to fall to the Taliban since 2001, when the group was ousted from government, earning the group perhaps its greatest victory since the US-led invasion. Around 500 Taliban fighters entered and seized control of the city on 28 September, one day before the first anniversary of the national unity government.

Abdullah subsequently declared that the Taliban had 'infiltrated the security structure and started the operation from within the city'. A spokesman for the Ministry of Defence blamed the incident on a lack of coordination among security forces, while other officials suggested that the ANSF had known about the attack, but was unable to adequately prepare for it. Ghani quickly dismissed Kunduz Governor Mohammad Omar Safi, who had been out of the country at the time of the battle.

During a series of offensives in April and June, the Taliban had gained control of much of the countryside surrounding Kunduz. This enabled a force build-up in several districts, which were used as launching pads for the capture of the city. The assault came from the south, the southwest and the north, blocking off entry points for government reinforcements, which were further delayed by ambushes and attacks using improvised explosive devices.

Nonetheless, US airstrikes played a significant role in the battle, and likely prevented the fall of the city's airport. One such operation had a tragic outcome: on 3 October, following a request for air support from Afghan forces, a US AC-130 *Spooky* gunship fired on the main hospital in Kunduz. Médecins Sans Frontières, which ran the facility, reported that the attack killed 42 people, including patients and members of its staff. Obama apologised for the assault, while a military inquiry concluded that it resulted from an 'avoidable human error'.

The capture of Kunduz projected the image of a strong, united Taliban, and helped Mansour rebut the reports of discord within the group that had persisted since the announcement of Mullah Omar's death. The new leader stated that 'people who said we were a small force with an unchosen leader can now see how wrong they were about the potential and strength my people have'.

The UN declared that around 289 civilians were killed and 559 others injured during the battle for Kunduz. According to Kabul, 20,000 families, or more than 100,000 people, were displaced by the fighting. Mansour said the Taliban had 'no intention of transgressing against [citizens'] personal property, carrying out extrajudicial killings, looting or breaching the inviolability of homes'. But the UN stated that the insurgents had carried out 'extrajudicial executions, including of healthcare workers, abductions, denial of medical care and restrictions on movement out of the city'. The Taliban also released hundreds of inmates of Kunduz prison, many of whom were members of the group, and reportedly looted 'vast quantities' of weapons and vehicles.

The Taliban announced its withdrawal from the city on 13 October. This followed a pattern evident in other offensives in the year: the insurgents would overrun an urban area, which government forces would retake shortly afterwards. While these victories undermined confidence in the unstable government in Kabul, they also suggested that the Taliban was unable to hold territory.

Helmand Province, in southern Afghanistan, came under sustained attack from the Taliban throughout the year. Musa Qala district fell to the insurgents in late August, after heavy fighting. The US deemed the battle there important enough to warrant a large number of airstrikes in support of the ANSF, which was nonetheless forced on 18 August to retreat from the centre of Now Zad (while claiming to have retained control of the district). In October, civilians began to flee Lashkar Gah, the provincial capital, as the Taliban attempted to seize control of surrounding areas – employing the same tactic used in its successful operation in Kunduz.

NATO was crucial to holding back the Taliban advance: both the US and the United Kingdom deployed special forces to Helmand in response to the intense fighting, and American troops played a central role in preventing vital areas falling to the insurgents. Yet on 13 December, after a month of fighting, the Taliban seized control of all of Marjah district except for the governor's compound and the headquarters of the security forces. Issuing a public plea to President Ghani on 20 December, Helmand Deputy Governor Mohammad Jan Rasoulyar warned that the entire province could fall unless 'urgent action' was taken. Rasoulyar appeared to have taken this step after Kabul denied his requests for military support, but it cost him his job.

On 8 December, around 16 Taliban fighters attacked the military base at Kandahar airfield, barricading themselves inside a school building near the facility before being killed in a clearing operation. The assault led to the deaths of 54 people, including 39 civilians, four of them children. Three days later, Taliban suicide bombers attacked a guest house belonging to the Spanish Embassy in a high-security area of Kabul, killing two Spanish police officers, four Afghan police officers and one civilian. On 21 December, another Taliban suicide bomber killed six US soldiers on patrol near Bagram air base in Parwan Province, the heaviest loss for US forces since 2012.

Delayed peace process

Recognising that Pakistan had a key role to play in Afghanistan's peace process, Ghani invested significant political capital in reaching out to Islamabad, a move that was deeply unpopular at home. On 10 February, he took the symbolically important step of suspending a request for weapons from India that had been made by his predecessor, Hamid Karzai. Afghan army cadets also began training in Pakistan for the first time. Pakistani Foreign Secretary Aizaz Ahmad Chaudhry stated in April that the level of military, intelligence and political contact between Kabul and Islamabad was 'unprecedented'.

During a May visit to Kabul by Nawaz Sharif and General Raheel Sharif, Pakistan's prime minister and army chief respectively, Afghanistan's intelligence service, the National Directorate of Security (NDS), signed a Memorandum of Understanding with its Pakistani counterpart, Inter-Services Intelligence (ISI). The deal reportedly centred on intelligence sharing, anti-terrorism cooperation and the ISI's provision of training and equipment to NDS personnel. But the arrangement caused an uproar in Kabul, forcing Ghani to downplay its significance and eventually to suspend its implementation. Pakistan appeared to provide little in return for his efforts beyond supportive statements. Attempting to quell public discontent, Ghani stated on 25 May that Pakistan was engaged in an 'undeclared war' with Afghanistan. Five days later, the press published a letter to Pakistani leaders in which Ghani demanded that they demonstrate sincere support for the peace process.

Prior to the controversy over the deal, the peace process appeared to be making some progress. Informal discussions on the issue were held at a May conference in Qatar, which was attended by representatives of the UN, the Taliban, the Afghan and Pakistani governments, and Hizb-e-Islami – led by former warlord Gulbuddin Hekmatyar. The parties agreed that Afghanistan's political system and constitution 'should be discussed in detail', and the Taliban indicated that it could support a power-sharing deal. Although little was achieved in a series of semi-official meetings involving the Taliban and Afghan representatives in Oslo and Dubai in June, these sessions were followed on 7 July by the first official talks between Kabul and the Taliban, held in Murree, in Pakistan. The Taliban's political office in Qatar dismissed the talks in Murree as 'unauthorised', claiming that the ISI had 'hijacked' the Taliban representatives there. In Murree, the Afghan government pushed unsuccessfully for a ceasefire, and the discussions were set to continue on 31 July. However, the process was suspended by the announcement of Mullah Omar's death (according to the NDS, he died in April 2013 in a Karachi hospital; Pakistan's authorities denied he was ever present in the country). Coinciding with the breakdown in relations between Kabul and Islamabad, the declaration created disunity within the Taliban, and prompted Islamabad to cancel the second round of official talks.

Mansour was thought to favour negotiations with Kabul, and to have sanctioned the meeting in Murree. Nonetheless, he stated on 1 August that 'we will continue our jihad until the creation of an Islamic system' and reiterated his support for Mullah Omar's mission, describing the peace process as 'enemy propaganda'. It remained unclear whether Mullah Omar had backed peace efforts, but a 14 July statement issued in his name – likely authored by Mansour or his associates – supported negotiations with the government in principle.

The peace process only began to regain momentum in early December, when the Heart of Asia conference was held in Islamabad, providing an opportunity for rapprochement between Afghanistan and Pakistan. In the lead-up to the event, Ghani said there was 'considerable uncertainty whether Pakistan would truly acknowledge a sovereign Afghan state', but added that Prime Minister Sharif had gone 'a very long way' to reassure Kabul on the issue. Ghani also hedged his approach to Pakistan by taking a more positive stance on India. This shift became particularly apparent on 25 December, when Indian Prime Minister Narendra Modi visited Kabul to inaugurate a new parliament building constructed with Indian financial assistance. Furthermore, New Delhi agreed to donate four Mi-25 attack helicopters to Afghan forces.

After relations between Kabul and Islamabad began to warm again, Pakistan and the US agreed to resume their pursuit of a dialogue with the Taliban. China also expressed its support for the peace process, engaging considerably more with Afghanistan in 2015 than it had for many years. It emerged that a Taliban delegation had visited China in late 2014, and Afghan Deputy Foreign Minister Hekmat Karzai stated in March that Kabul had asked Beijing to pressure Islamabad to support the peace process. China's renewed interest stemmed partly from concern about militancy in Xinjiang, its western province bordering Afghanistan, as well as investments in its 'One Belt, One Road' regional economic initiative.

ISIS and a divided Taliban

Mansour was acting head of the Taliban for many years before being named as Mullah Omar's successor. He selected Sirajuddin Haqqani and Haibatullah

Akhunzada as his deputies, with the former also serving as head of operations. Both members of the Haqqani network, Sirajuddin Haqqani was the son of its late founder, Jalaluddin Haqqani, while Akhunzada had run the Taliban's courts system. Soon after Mansour announced the appointments, there were reports of infighting among the Taliban leadership and between commanders on the ground, due to disagreements over the secrecy with which he had been chosen to lead the group. Mullah Omar, whom all Taliban fighters had sworn to obey, had served as a unifying figurehead in an often fractured movement. As such, many members of the leadership and insurgents on the front-line felt betrayed by the concealment of his death, and some sought to join local ISIS affiliate Khorasan Province (ISIS–KP).

Opposition to Mansour initially centred on Mullah Mohammad Yaqoub and Abdul Manan, Mullah Omar's son and brother respectively, as well as former Taliban military commander Mullah Abdul Qayum Zakir. Yaqoub and Manan eventually agreed to support Mansour, as did al-Qaeda leader Ayman al-Zawahiri. But Syed Tayyab Agha, head of the political office in Qatar, stated that he would 'not support any side in the current internal disputes within the Taliban' and resigned from his post, arguing that the concealment of Mullah Omar's death had been a 'historical mistake'.

On 7 August, the Taliban carried out three coordinated suicide attacks in Kabul, killing more than 50 people and injuring more than 400 others. This was the highest number of civilian casualties in a single terrorist operation since 2009, according to the UN Assistance Mission in Afghanistan. Taliban spokesman Zabihullah Mujahid said the objective of the attacks was to 'convey a message to those saying the Taliban had been split into factions'. The group carried out three more suicide attacks in Kabul in August, killing around 20 civilians and injuring more than 80 others. Each of the operations helped strengthen Mansour's position, as did the capture of Kunduz.

Nonetheless, the Taliban's internal divisions widened further in November, with Mullah Mohammad Rasool Akhund named as leader of the main faction opposing Mansour's leadership. A week of battles between rival Taliban groups in Zabul Province led to the deaths of dozens of militants, some of the heaviest losses the group had ever incurred in such clashes. On 11 November, there were unconfirmed reports that Mullah Mansour Dadullah, Rasool's most powerful commander, had been killed by fighters loyal to Mansour. The following month, the Taliban was forced to deny rumours that Mansour had been injured or killed in a shoot-out with another commander in the Pakistani city of Quetta. To this end, the group released on 5 December an audio message ostensibly from Mansour; although its authenticity was not verified, the recording referred to recent events.

The infighting was exacerbated by the emergence of ISIS–KP, which had created a new dynamic in the Afghan conflict. The central leadership of ISIS had accepted the group as its local affiliate in January, appointing Hafiz Saeed Khan, a former member of Tehrik-e-Taliban Pakistan (TTP), as the emir of Afghanistan, Pakistan and parts of neighbouring countries. It was thought that most members of ISIS–KP were Pakistani militants pushed out of Waziristan by *Zarb-e-Azb* – Islamabad's military operation against the TTP – as well as disaffected members of the Afghan Taliban and other Afghan recruits. For Mansour's group, ISIS–KP was a new rival for legitimacy, territory and recruits. While the Afghan or Pashtun nationalists of the Taliban sought to control Afghanistan, ISIS aimed to expand its transnational 'caliphate' far beyond the Middle East and South Asia. Thus, in a letter published on 16 June, Mansour told ISIS leader Abu Bakr al-Baghdadi that jihadist activities in the country were under the sole leadership of the Taliban, which would be 'forced to act' if ISIS interfered there. Tensions between the groups increased further when the central leadership of the Islamic Movement of Uzbekistan, once an ally of the Taliban, pledged allegiance to ISIS days after the confirmation of Omar's death.

A September report by the UN claimed that ISIS–KP was active in 25 of Afghanistan's 34 provinces. Many of the group's fighters were concentrated in the eastern province of Nangarhar, where they contested several districts, and established a stronghold in Achin district. It also has a small presence in southern provinces such as Ghazni and Zabul, as well as Kunar and Nuristan, in the northeast. The first major attack by ISIS–KP took place on 18 April, when a suicide bomber killed 35 people and injured 125 others outside a bank in Jalalabad, in Nangarhar.

The group contested much of Nangarhar throughout 2015, despite being regularly targeted in strikes by US unmanned aerial vehicles and the Afghan Air Force, as well as ground operations by both international troops and the Taliban – the latter of which formed a special unit of more than 1,000 fighters to combat the new threat. In one of these strikes, conducted on 9 February using a US unmanned aerial vehicle in Kajaki, in Helmand Province, ISIS–KP deputy leader Mullah Abdul Rauf Khadim was killed along with five other militants. The NDS announced on 1 July that it had also established a special force to combat ISIS–KP, which reportedly recruited 700–800 fighters by the end of the month. The ANSF launched on 1 August a month-long operation against the Taliban and ISIS–KP in three Nangarhar districts. This was the Afghan security forces' first major offensive against ISIS–KP, having previously appeared to leave the fight to the Taliban.

As in other parts of the world, ISIS imposed strict laws in the areas it controlled. The group reportedly formed its own judiciary in some parts of Nangarhar, requiring locals to refer to the new body in resolving disputes. The jihadists were also said to have closed schools and prohibited women from leaving their homes without a burqa and a 'Sharia excuse', while running prisons whose inmates included tribal elders, religious scholars and members of the security forces. Furthermore, ISIS–KP burned poppy fields in territory captured from the Taliban, depriving both its rival and local farmers of income. On 11 August, ISIS released a video in which its fighters used explosives to execute ten people in Nangarhar Province, mirroring its brutal conduct in Iraq and Syria. The Taliban condemned the killings as 'horrific'. Hajji Ghalib, governor of Achin, claimed in September that ISIS–KP had killed 600 people in the district in the preceding four months.

General John Campbell, commander of *Operation Resolute Support*, stated in December that some foreign fighters from Syria and Iraq had joined ISIS–KP in

Nangarhar, and that there were indications that it was attempting to strengthen its links with the central leadership. The group also stepped up its propaganda efforts in the province, broadcasting daily from an FM radio station called 'Voice of the Caliphate'.

In response to the increasing insecurity in Afghanistan, US Secretary of Defense Ashton Carter requested in December broader authority to expand the fight against ISIS–KP from the White House. Meanwhile, Moscow stepped up its military engagement with Kabul: Zamir Kabulov, Russian special envoy to Afghanistan, stated that the country's requests for weapons would be carefully considered, and that Russia would supply 10,000 AK-47s to Afghan forces in January 2016. Kabulov also said that Moscow was sharing information with the Taliban to aid the fight against ISIS–KP, although Mansour's group denied this.

Al-Qaeda also retained a presence in Afghanistan. Abu Khalil al-Sudani, according to the Pentagon head of al-Qaeda's suicide and explosives operations, was killed on 11 July by a US airstrike in Bermal district, in Paktika Province. In October, more than 200 US and Afghan personnel conducted a five-day ground assault in Kandahar's Shorabak district, targeting what Gen. Campbell described as 'probably the largest' al-Qaeda training camp ever found in Afghanistan. The operation involved 63 airstrikes by American aircraft. Later that month, the US used an unmanned aerial vehicle to kill al-Qaeda commander Bilal al-Taib and two of his subordinates in Ghaziabad district, in Kunar Province.

Civilians exploited and displaced

The intensification of the conflict had a severe impact on Afghan civilians, around 3,500 of whom were killed during the year, according to the UN. However, the dynamics of the humanitarian situation remained the same in most of Afghanistan, albeit with ISIS–KP creating a new, particularly violent threat to civilian life. The UN estimated that insurgents and pro-government forces were solely responsible for 62% and 17% of civilian casualties, respectively, and attributed 17% to crossfire between the sides. The Taliban killed many civilians by conducting suicide bombings in population centres, targeting government officials and buildings, as well as foreigners. In one such attack, carried out on 2 April, 20 people were killed and 64 others injured at a protest against the acting governor in Khost (the Taliban denied any involvement in the assault, as it often had when large numbers of civilians were killed).

The UN reported the Taliban had increased the 'scale and depth' of its involvement in economic crime, including illegal mining, taxation, kidnapping for ransom, and drug production and trafficking. Such activities were an increasingly important source of revenue for the group. In its annual Afghanistan Opium Survey, published in October, the UN Office on Drugs and Crime found that cultivation of poppies for opium fell by 19% in 2015, but that the number of poppy-free provinces also declined.

The UN also noted that the insurgents were increasingly targeting aid workers, who were often killed or kidnapped. In one such incident, five employees of Save the Children were abducted in Uruzgan Province on 3 March, and their bodies discovered on 10 April. The charity subsequently suspended all of its activities in the province. Insurgent and criminal groups also continued to kidnap and ransom Afghan civilians, particularly members of minorities. The Islamic Movement of Uzbekistan abducted around 31 ethnic Hazaras – who are predominantly Shia, and as such often targeted in sectarian violence – in Zabul Province on 23 February. Nineteen of the victims were released in May, and another in November. On 8 November, the bodies of seven ethnic Hazaras kidnapped in Ghazni the previous month were found in Shah Joy district, in Zabul. The killings sparked large-scale protests across Afghanistan, with demonstrators demanding that the government protect the vulnerable ethnic and religious minority. Yet, two weeks later, unidentified militants kidnapped 16–30 ethnic Hazaras in Zabul.

The Afghanistan Independent Human Rights Commission stated in December that 1.2 million Afghans were displaced in the year, most of them by the conflict. Afghan parliamentarians called on the government to take action to prevent brain drain, as thousands of young people fled the country. Indeed,

Afghans made up one of the largest groups among the refugees travelling to Europe from the Middle East, South Asia and Africa. Meanwhile, Pakistan declared that it would begin to repatriate 1.5m Afghan refugees in its territory. Islamabad had long viewed the refugees as a burden and a threat to national security.

India (Assam)

Key statistics	2014	2015
Conflict intensity:	Low	Low
Fatalities:	300	60
New IDPs:		
New refugees:		

The security and political situation in Assam improved markedly in 2015, with fewer insurgency-related fatalities in the year than in the last three months of 2014. This was partly due to the steady progress made by the peace process there, as well as similar initiatives in neighbouring states, such as Nagaland. While hardline armed groups that opposed the peace negotiations sought to coordinate their activities – and thereby strengthen their military capabilities – they came under severe pressure from sustained counter-insurgency operations.

The Congress-led government of Assam Chief Minister Tarun Gogoi underwent a reshuffle in January 2015, making 14 appointments to ministerial positions. The episode marked the end of an eight-year coalition between Congress and the Bodoland People's Front, with Chandan Brahma, chairman of the latter party, withdrawing his support for the government and creating a vacancy in the cabinet.

In January, the Supreme Court of India criticised Assam's government for 'dragging its feet' on an issue that had shaped the conflict in the state since 1979: the legislature's response to the influx of illegal migrants from Bangladesh. Congress had initiated in September 2014 a scheme for receiving as many as

6,690,000 applications for citizenship from residents who were on either the 1951 or 1971 electoral rolls, and the process for implementing this initiative continued throughout 2015 as the state government was criticised by a Supreme Court commission for the slow pace of implementation of border fencing and patrolling initiatives. Assam's government urged New Delhi to complete the construction of fences on the border with Bangladesh in the first few months of 2015, to no avail. According to Union Minister of Home Affairs Rajnath Singh, the project would be completed in 2016 after it remained unfinished at the end of 2015.

In early 2015, the Indian government reportedly ruled out further peace talks with new armed groups, although it reiterated its willingness to pursue negotiations with existing armed groups – the most significant of which were the National Democratic Front of Bodoland–Progressive (NDFB–P) and the pro-talks faction of the United Liberation Front of Asom (ULFA–PTF).

Negotiations with ULFA–PTF did not produce an accord or ceasefire agreement during the year, but instead involved a series of optimistic statements from both sides, as well as gestures with potentially positive effects. As part of the negotiations, the government granted scheduled tribe status to six communities: the Koch-Rajbongshis, the Tai Ahoms, the Morans, the Motoks, the Chutiyas and the Tea. On 11 November, New Delhi completed negotiations with Dhaka to transfer ULFA General Secretary Anup Chetia into Indian custody. On 24 December, after it had become clear that Chetia was willing to lead the ULFA–PTF in joining the peace process, the Indian government allowed him to be released from jail. This was a significant boost to the peace process, and it dealt a blow to the United Liberation Front of Asom–Independent (ULFA–I): Paresh Barua, the group's leader, became the only founding member of the ULFA not involved in the peace process. Meanwhile, on 9 October, the NDFB–P began peace negotiations with the government, although it reiterated its demand for an independent Bodo state.

There continued to be little prospect of peace talks with groups conducting attacks in Assam, such as the ULFA–I and the National Democratic Front of

Bodoland–Songbijit (NDFB–S), particularly after Union Minister of State for Home Affairs Kiren Rijiju ruled out such moves due to their participation in 'mindless violence'. Nonetheless, the ULFA–I reiterated its willingness to negotiate with the Indian government, on the conditions that this occurred in a neutral country under UN supervision, and included a discussion of sovereignty.

Counter-insurgency operations against hardliners

The Indian Army continued *Operation All Out* against NDFB–S throughout 2015, arresting 365 of the group's fighters (the bulk of them in the first five months of the year) and killing at least 19 others. Given that the NDFB–S had been estimated to have only 270 members, this suggested that it was severely degraded by the counter-insurgency campaign. *Operation All Out* was extended to Chirang district on 18 September, reportedly forcing the group's leader, I.K. Songbijit, to flee to Myanmar, while other senior figures escaped to Bhutan.

The security forces also targeted and significantly weakened, inter alia, the Karbi People's Liberation Tigers, arresting the group's senior leadership and 60% of its personnel in February – a move that cast doubt on its capacity to continue operations. The counter-insurgency effort also allowed many of those displaced by violence in December 2014 to return to their homes, after which the government provided support packages to help them rebuild. Despite conducting these aggressive campaigns, the security forces suffered just one fatality in Assam in the year. Nonetheless, some civilians were trapped between security operations – one of which led to the death of a resident of Karbi Anglong district, due to mistaken identity – and insurgent groups increasingly focused on extortion and kidnapping.

Conducting a series of low-profile attacks throughout the year, the ULFA–I appeared intent on re-establishing a presence in Assam, having been absent since splitting with the ULFA–PTF. These included an extortion-related grenade assault on 13 February, which killed two people, and an attack at a stone-crusher unit in Namrup on 30 November, which critically injured two Hindi-speakers. According to Indian intelligence, the resort to violent extortion by both the

ULFA–I and the NDFB–S reflected the financial problems they faced due to the counter-insurgency operations.

Assam's security forces and politicians were concerned by the formation in April of regional umbrella organisation the United National Liberation Front of Western South East Asia, which was headed by the National Socialist Council of Nagalim–Khaplang (NSCN–K) and included the ULFA–I, the NDFB–S and the Kamtapur Liberation Organisation. Yet the organisation had no considerable operational success in the state, and its formation appeared to have been spurred by the increasing weakness of its members. Barua's claim that the addition of seven more groups from Manipur would 'tip the balance' against the security forces in Northeast India stood in stark contrast to a year in which, overall, Assam's peace process made substantial progress, albeit without achieving conclusive results.

India (Manipur)

Key statistics	2014	2015
Conflict intensity:	Low	Low
Fatalities:	55	100
New IDPs:		
New refugees:		

The political climate in Manipur was shaped by instability, protest and continuing unrest over demands for an Inner Line Permit (ILP) system, designed to restrict access and residential rights in the state for non-Manipuri Indians and foreigners. Accordingly, the legislature passed on 16 March 2015 the Manipur Regulation of Visitors, Tenants and Migrant Workers Bill to implement compulsory registration for these groups of people. But the version of the bill that was passed drew criticism from opposition groups such as the Bharatiya Janata Party (BJP), which regarded it as a compromise on the initial draft. The move was also opposed by the body that submitted the bill, the Joint Committee on Inner Line Permit

System, due to its failure to limit the length of time that outsiders could remain in Manipur. Protesting this shortfall, the committee held an 18-hour strike on 15 March, while the Manipur People's Party staged a sit-in at its party office on 24 March.

In July, the protests against the ILP became particularly violent, causing the death of one person and the injury of 300 more. Due to the political pressure this created, the bill was withdrawn. In a bid to rectify the situation, the legislature introduced on 26 August the Protection of Manipur People's Bill, 2015, the Manipur Land Revenue and Land Reforms (Seventh Amendment) Bill, 2015 and the Manipur Shops and Establishments (Second Amendment) Bill, 2015. However, the measures did little to quell the unrest, which resulted in another eight fatalities and the imposition on 1 September of a curfew in Churachandpur district, with protesters claiming that their land was being given to 'outsiders'.

The state had also experienced political unrest on 27 April, when the United Naga Council held a 48-hour *bandh* (strike) in areas controlled by the state's Nagas, to demand that the Indian government provide them with an 'alternative arrangement' outside of the Government of Manipur, in advance of a broader Indo-Naga political settlement.

Among the roughly 40 insurgent groups operating in Manipur, there were several clashes between the two largest factions of the National Socialist Council of Nagalim (NSCN): Isak Muivah (NSCN–IM) and Khaplang (NSCN–K). Five NSCN–K militants were killed by the NSCN–IM on 22 May, and there were reports of another battle between the factions in August, after the bodies of three NSCN–K fighters were found in Senapati district. The NSCN–IM and the Zeliangrong United Front (ZUF) clashed over control of Tamenglong district throughout 2015, with the ZUF repeatedly accusing the Indian government of aiding its opponent during the confrontations. These complaints were especially prominent on 1 March, when the ZUF alleged that the government had taken part in an NSCN–IM operation that killed one ZUF militant. Later that month, battles between the ZUF and the Indian military took place in Tamenglong district, injuring seven army personnel. The conflict between various parties in

Tamenglong continued to affect civilians living there. On 21 February, as each group sought to assert its authority, the ZUF shot and critically injured a local woman, sparking a wave of violence that resulted in the lynching of a ZUF militant on the same day. Meanwhile, the NSCN–K suffered a blow in the district, as its Tamenglong-based Zeliangrong faction left to form the NSCN–Reformation (NSCN–R), which favoured talks with the government.

There were also fratricidal battles between the various Kuki factions in the year: the bodies of two Kuki National Army (KNA) militants were discovered on 19 January in Churachandpur district, following a suspected clash with the Kuki Independent Army. And, in August, the KNA and the Kuki National Front engaged in a spate of retaliatory killings in Thoubal and Churachandpur districts.

Repercussions from Nagaland

The peace process in Nagaland continued to have repercussions for Manipur, due to the presence of the NSCN–IM and the NSCN–K in the latter state. While the NSCN–IM was engaged in peace talks throughout 2015, concluding a framework accord on 3 August, the deal was treated with suspicion by Manipur Chief Minister Okram Ibobi Singh, who claimed that he had not been consulted during the process. Given that the NSCN–IM laid claim to the Naga-dominated Manipuri districts of Ukhrul and Senapati as part of 'Greater Nagaland', which had resulted in major clashes between Nagas and Kukis in preceding years, the accord also elicited a hostile response from Manipur's Kuki and Meitei populations.

There was a significant setback in the peace process between the Indian government and the NSCN–K on 31 March, when the group announced that it would not extend the sides' 14-year ceasefire without being granted a discussion on Naga sovereignty; the government responded by declaring that the ceasefire was no longer valid. Furthermore, the NSCN–K laid an ambush in Tamenglong during the month, killing four members of the Gurkha Rifles in an apparent demonstration of its willingness to return to violence. The NSCN–K also announced in April the establishment of regional umbrella organisation the United National Liberation Front of Western South East Asia (UNLFWSEA), which was designed to strengthen military cooperation between NSCN–K, its allies in the Coordination Committee – a conglomeration of Manipuri insurgent groups – and Assam-based rebel organisations such as the United Liberation Front of Asom–Independent (ULFA–I) and the National Democratic Front of Bodoland–Songbijit.

Manipur's most high-profile security incident of the year occurred on 4 June, when the NSCN–K launched an attack on a 6th Dogra Regiment convoy in Chandel district, resulting in the deaths of 18 soldiers and two militants. The group subsequently announced that the assault, one of the largest and most deadly strikes on the Indian Army in years, had involved the assistance of the Kanglei Yawol Kanna Lup, a member of the Coordination Committee. The ULFA–I, which also provided assistance, later lauded the operation as a major success for UNLFWSEA. New Delhi's response to the attacks largely took place outside of Manipur, with the air force killing 15 militants in surgical strikes in Myanmar shortly thereafter, while coordinating with the country's government and armed forces to enable follow-up operations, though it is unclear whether further Indian surgical operations have taken place since the Myanmar Army moved its troops to the border with India in June. Fatalities among the security forces drastically declined after the 4 June assault, and overall fatalities also fell in the second half of 2015, likely the result of both the cross-border developments and the intensification of counter-insurgency operations within India.

Meanwhile, the People's Liberation Army, the military wing of the Revolutionary People's Front, was one of several other militant organisations to remain active in Manipur throughout 2015, killing two personnel from the Assam Rifles on 1 April and injuring two members of the security forces on 1 June. The counter-insurgency operations in Myanmar also affected the group: the Indian Army killed three of its fighters during a raid near Chandel district on 5 September, in a move that drew protests from Naypyidaw. This friction between Myanmar and India suggested that their tacit understanding on cross-border operations was incomplete.

Pervasive insecurity

Residents of Manipur faced considerable risks from all militant factions in the year, with 27 civilian deaths and 46 recorded cases of extortion related to insurgent or counter-insurgency activity. Due to the competition between armed groups in the region, such extortion was often violent, and sometimes involved abductions and murders. On 7 February, locals found the body of a civilian kidnapped two days earlier by the ZUF, and, on 12 May, two more civilians were killed after being abducted by the Kuki Revolutionary Front. This insecurity permeated the politics of the state, particularly in the weeks prior to the 1 June Autonomous District Council elections: a Congress party worker was kidnapped and murdered by unidentified militants in Tamenglong on 6 May; the wife of a BJP worker was kidnapped from Chandel district by unidentified militants on 24 May; and the chairman of Taninglang village, in Ukhrul district, was abducted by the NSCN–IM on 30 May. The day before the election, militants attempted to kidnap polling officials and steal ballot boxes. Civilians were also affected by the actions of the security forces, both in encounters involving insurgents and in civil unrest. Throughout the year, Manipur was destabilised by the interlinked violence associated with the ILP, the insurgency and counter-insurgency operations, as well as the internecine fighting between armed groups.

India (Nagaland)

The peace process between the Indian government and the largest armed group in Nagaland, the National Socialist Council of Nagalim–Isak Muivah (NSCN–IM), slowly progressed in the first half of 2015. New Delhi's representative in the negotiations, R.N. Ravi, stated in May that the issues of sovereignty and

Key statistics	2014	2015
Conflict intensity:	Low	Low
Fatalities:	15	100
New IDPs:		
New refugees:		

integration were on the agenda, but added that there was no possibility of outright independence for the Nagas. Nonetheless, the sides overcame their differences to sign a framework agreement on 3 August.

The document was designed to form the basis of a more comprehensive peace deal, scheduled to follow within three months. In the meantime, governor of Nagaland and Assam P.B. Acharya announced on 9 September that as many as 4,000 NSCN–IM militants would be absorbed into the security forces, thereby demonstrating the authorities' commitment to rehabilitating fighters. Although the sides failed to reach a comprehensive deal in the year, Ravi held talks in December to encourage other Naga political and militant groups to sign the framework document, having pressured them to do so on 21 November by declaring that the government would not allow separate agreements for separate groups. Significantly, the framework agreement allowed the NSCN–IM to claim 'shared sovereignty' over Naga-inhabited areas with the Indian state, although the exact territorial demarcations remain ambiguous. This provision allowed the group to match the rhetoric of organisations such as the Naga National Council (NNC), which first declared Naga independence in 1947. The NSCN–IM also used the framework agreement to assert its leadership of the Naga political movement, claiming that the NNC had forfeited the right to continue the Naga struggle by signing the 1975 Shillong Accord.

Despite the NSCM–IM's adherence to the deal, there were around 100 fatalities in Nagaland involving Naga armed groups in 2015, as many as in the previous three years combined, and more than in any year since 2008 (when 200 people were killed). This was in part due to the expiration of the 14-year ceasefire agreement between the Indian government and the National Socialist Council of Nagalim–Khaplang (NSCN–K), which followed that group's declaration that it would only extend the accord if granted a discussion on Naga sovereignty. Yet there was evidence of a disagreement within the NSCN–K over the declaration: S.S. Khaplang, the group's chairman, expelled two ministers who were in favour of extending the ceasefire. This resulted in the second recent split in the NSCN–K, with the formation of the pro-talks National Socialist Council of Nagalim–Reformation (NSCN–R), led by the two ministers and largely comprised of Nagas from the Tangshang and Konyak tribes. The new group signed a ceasefire agreement with the Indian government on 28 April, although it simultaneously sought to establish its presence through acts of extortion against civilians both within Nagaland and in Arunachal Pradesh. However, there was a suggestion that the split had failed to create unity of purpose in the NSCN–K, with its announcement in September that it would not sign Myanmar's nationwide ceasefire agreement due to 'political issues within the group'.

During the first three months of 2015, the NSCN–K coordinated some attacks outside Nagaland – such as the killing of an Assam Rifles trooper in Changlang district, in Arunachal Pradesh, on 6 February – but the expiry of the ceasefire with the Indian government led to a significant escalation in violence. The NSCN–K killed three soldiers during an ambush in Arunachal Pradesh in April, and, on 3 May, the group gunned down eight members of the security forces in three different assaults. The violence culminated on 4 June in one of the largest and most deadly attacks on the Indian Army in years: an ambush on a 6th Dogra Regiment convoy in Manipur that killed 18 soldiers and injured 11 others. The NSCN–K subsequently announced that it had carried out the assault with the support of Manipuri groups Kanglei Yawol Kanna Lup (KYKL) and the Kangleipak Communist Party (KCP), as well as Assamese group the United Liberation Front of Asom (ULFA). The attack was presented by NSCN–K and other regional armed groups as a victory for the United National Liberation Front of Western South East Asia (UNLFWSEA), an umbrella organisation they had formed in April to strengthen their cooperative efforts.

Effects of the insurgency

New Delhi responded to the 4 June attack by launching a sustained counter-insurgency operation against the NSCN–K, quickly conducting surgical strikes on militant camps in Myanmar that killed 15 suspected fighters from the group and the KYKL. The Indian government also asked Naypyidaw to capture and extradite the NSCN–K leadership residing within Myanmar, while broadening security cooperation with its neighbour and obtaining assurances that the Tatmadaw would not prevent Indian security forces from conducting cross-border strikes against the militants. The Tatmadaw also deployed extra brigades near Myanmar's border with India, applying greater pressure on the NSCN–K.

As the Naga-inhabited areas near the border were a major base of operations for the NSCN–K, the counter-insurgency measures likely reduced its capacity to carry out large-scale attacks. Indeed, the group did not kill any more members of the security forces in the remainder of 2015 – although, on 4 October, it executed two labourers accused of spying for the government. Furthermore, in an effort to strengthen its position as leader of the UNLFWSEA, the NSCN–K provided assistance to groups such as the ULFA and the Meghalaya-based Liberation of Achik Elite Force.

Despite the fact that some progress towards peace was made in the year, Nagaland's armed groups – including pro-talks factions such as the NSCN–IM and the NSCN–R – continued to carry out human-rights abuses against civilians. Both the NSCN–IM and the NSCN–R freely engaged in extortion, with the former group claiming that it had 'every right' to do so. Members of the Assam Rifles claimed that the NSCN–K continued to forcibly recruit young people, nine of whom were kidnapped by the group in Phek district. Such activities exacerbated concern about the use of child soldiers in Nagaland. The NSCN–K was also reported to have smuggled improvised explosive devices into Nagaland, with the intent of using them against both the security services and civilians. This was reflected in several bombings targeting civilians in Dimapur during the year, including an attack on 5 January that critically injured one, and an assault on 27 June that killed two others.

Inter-tribal clashes also continued to threaten civilians. On 7 February, one such battle in Tuensang district between the Chang and Yimchungru tribes resulted in the displacement of 1,000 locals. The violence there also led to the police firing on Chang protesters, after the discovery on 14 February of the body of a Chang teenage boy who had been abducted that day provoked anger among the Chang community, resulting in the ransacking of the deputy commissioner's office in the district. Meanwhile, Amnesty International raised concerns about flaws in the justice system, following an incident on 5 March in which a suspected rapist was abducted from prison and lynched, amid claims that he was an illegal immigrant from Bangladesh. The pervasive insecurity affecting Nagaland's civilians was exacerbated by a series of battles between armed groups, including a clash on 29 April between the NSCN–IM and the Zeliangrong United Front, as well as two attacks carried out on 30 November and 19 December by the National Socialist Council of Nagalim–Unification on the NSCN–R.

India (Naxalites)

Key statistics	2014	2015
Conflict intensity:	Low	Low
Fatalities:	300	300
New IDPs:		
New refugees:		

In 2015, there was a moderate but steady decline in violence in India's Red Corridor – the key area of operation for communist militants the Naxalites. As fractionalisation within the movement helped reduce the threat it posed, state-level politicians declared that the Naxalites faced total defeat in several states and districts.

Large numbers of communist militants continued to surrender to law-enforcement agencies. Fighting between the Communist Party of India–Maoist (CPI–M) and the security forces appeared to heavily favour the latter.

Despite its continuous loss of fighters and territory, the CPI–M still presented a serious challenge to the Indian authorities. As the group's influence waned in some areas, it began to conduct operations in previously uncontested states and districts. The group also seized on a new propaganda opportunity in the year, denouncing mining projects bitterly opposed by some local tribes. The CPI–M continued to effectively fund itself through the extortion of legitimate businesses, as well as by providing protection to smugglers and criminal groups. By the end of 2015, it was unclear whether the Indian security forces would capitalise on their battlefield advantage to prevent the CPI–M from securing new footholds across India.

Naxalism in retreat

The security forces' battlefield victories in many areas prompted Sajal Chakraborty, chief secretary of Jharkhand State, to announce in January that the Naxalites would be totally ousted from his state by April 2015. While this prediction proved overly optimistic, there was significant progress in Jharkhand. The Naxalites did not attack any police stations in the state during the year – for the first time since the movement began operations there. Nor did the CPI–M seize any police weapons. And, for the second year in a row, the group failed to convene a *Jan Adalat*, a kind of 'people's court' in which the militants summarily executed suspected informers and landlords accused of abuse. Twelve militants from Jharkhand surrendered to police during the year.

The security forces also reportedly made progress outside Jharkhand. By the end of the year, police in Chhattisgarh State declared Surguja district free of Naxalites. Chief Minister Raman Singh suggested that all of Bastar division, the traditional heartland of the Naxalite movement (comprising seven districts within Chhattisgarh), would soon be free of the insurgency. In Karnataka State's Udupi district, almost all eligible voters participated in district and local elections. Naveen Patnaik, Odisha's chief minister, reported in March that the Maoists' activities had waned in the state. Police reports from Kerala State suggested that the militants were no longer gaining new recruits from local tribes. And the annual People's Liberation Guerrilla Army week – a *bandh* (strike) held in honour of deceased CPI–M militants – was largely ignored by the civilian population. Combined, these developments suggested that support for the group was in decline.

Odisha state government reported that 136 militants had surrendered in 2015. Overall, there were at least five mass surrenders, involving at least 20 Naxalites each. Prominent Maoist leader Gajerla Ashok surrendered in Telangana State in December. While journalists uncovered hundreds of staged surrenders the previous year, there were no similar reports in 2015, suggesting that amnesty programmes had become more discerning and their statistics more reliable.

It seemed clear that the Naxalites' morale had been eroded. Nonetheless, the movement's various splinter groups continued to engage in internecine fighting. Militants from the Tritiya Prastuti Committee skirmished with their counterparts in both the CPI–M and the Jharkhand Prastuti Committee, in Bihar and Jharkhand states respectively. The CPI–M conducted a purge of its leadership in July, and subsequently claimed to have executed at least six influential leaders. Furthermore, a CPI–M spokesman admitted in December that former leader Sabyasachi Panda's 2012 decision to break away from the group continued to erode its internal unity (Panda, who formed the People's Liberation Front of India in 2012, was apprehended in 2014). This fighting within and between Naxalite groups served the interests of the Indian security forces.

Developing counter-insurgency efforts

Elements of the federal government began to press the security forces to integrate development projects into their strategic planning, modifying the 'security first' approach to the Naxalites advocated by Prime Minister Narendra Modi in 2014. As part of this, the Ministry of Home Affairs suggested in October that all police stations along the Red Corridor be designated as development centres. Given the existing policy of opening police stations in every 10–12 square kilometres of districts affected by the insurgency, the plan had the potential to build stronger ties between the state and thousands of remote communities. Furthermore, it was possible that a series of civil-engineering and medical projects proposed in New Delhi would significantly impede the Naxalites' recruitment efforts.

As well as integrating development projects into the remit of the security forces, some state and federal agencies also began to better familiarise themselves with local communities in the Red Corridor. The police superintendent in Maharashtra State's Gadchiroli district implemented in December a new rule requiring all his subordinates to learn basic Gondi, the dominant tribal language in the region. In Andhra Pradesh, the government announced that it would raise a battalion of tribal police officers. Endeavours such as these sought to create trust between the security forces and tribes, thereby improving the state's capacity to detect and defeat insurgents. However, some police officials cautioned that the proposed tribal battalion was reminiscent of the ill-fated, government-supported militia, which was banned in 2011 as unconstitutional, even as several thousand of its members were executed by the CPI–M.

Despite gains by the security forces, the CPI–M remained an effective, dangerous fighting force throughout the year. The group carried out a series of ambushes on police patrols, killing at least 50 officers. The two most damaging attacks occurred in Chhattisgarh. On 11 April, CPI–M militants operating in Sukma district killed seven officers from the state's Special Task Force, and injured 12 others, using an improvised explosive device (IED). Two days later, the group used an IED to destroy a mine-proof, ambush-protected vehicle in Dantewada district, killing five members of the Chhattisgarh Armed Force and wounding seven others. The Maoists also formed the Malkangiri–Visakhapatnam–Koraput division to increase their presence along the Odisha–Andhra Pradesh border. Despite perceiving an overall decline in Maoist activity in his state, Odisha Chief Minister Patnaik warned of an increasing insurgent threat in its peripheral districts. As a consequence, he requested in February that two additional battalions of the Central Reserve Police Force be deployed to deter the militants in Nuapada and Malkangiri districts.

The CPI–M continued to be a threat in many parts of its traditional heartland (especially the Bastar division), and Indian intelligence officials claimed in March that the group had begun to move into the Sahyadri mountain range, at the junction of Kerala, Karnataka and Tamil Nadu states. It was possible that the

Maoists had moved into this rugged terrain in search of greater protection from the security forces.

The CPI–M also remained a threat to state officials throughout the Red Corridor. Fearing that the Maoists would attempt to assassinate him, Telangana Chief Minister K. Chandrashekar Rao persuaded New Delhi to provide special equipment for protecting his official convoy from remotely detonated IEDs. The group also retained access to a great deal of funding. According to some estimates, the group collected around US$29 million per year by extorting mine owners, contractors and other traders, as well as businesses. Combined with its other sources of income, these funds were likely to ensure that it remained well equipped for some time.

The Naxalites attempted to compensate for their apparent loss of support among most civilians by taking advantage of aggrieved tribes. The group consistently opposed the distribution of mining licences in tribal areas, purportedly because mining often degraded the local environment. This damage, in turn, severely disrupted the cultural practices and livelihoods of tribes. Accordingly, the CPI–M assassinated several local political leaders who supported mining. In March, the group warned against plans to open an iron mine in Gadchiroli district. By appealing to anti-government sentiments among affected populations, the Maoists may gain access to a new source of recruits.

India–Pakistan (Kashmir)

Key statistics	2014	2015
Conflict intensity:	Low	Low
Fatalities:	175	200
New IDPs:		
New refugees:		

The conflict in Kashmir intensified in early 2015, as Indian and Pakistani border troops exchanged fire across the Line of Control. Clashes between the sides ebbed and flowed throughout the year, as did those involving local militant groups. India and Pakistan made several failed attempts to ease tensions through diplomacy, one of which led to a high-profile summit between the countries' national security advisers. Yet, despite these frustrated efforts, there was an unexpected thaw in relations late in the year, culminating in a surprise meeting in Lahore between Indian Prime Minister Narendra Modi and his Pakistani counterpart, Nawaz Sharif.

Unlikely coalition, bilateral tension

One of Jammu and Kashmir's first major political hurdles of 2015 related to legislative elections held in November and December the previous year, which had resulted in a hung parliament. Despite their hostility towards each other in the lead-up to the vote, Modi's Bharatiya Janata Party (BJP) and the Jammu and Kashmir People's Democratic Party (JKPDP), led by Mufti Mohammad Sayeed, entered into coalition negotiations in January. The process proved difficult due to the ideological differences between the parties: while the JKPDP advocated self-rule for Jammu and Kashmir and talks with local separatists, the BJP called for the state to be stripped of its autonomy. Nonetheless, they announced on 24 February that a coalition had been formed, with Sayeed appointed chief minister. The Jammu and Kashmir National Congress – which had lost control of the state parliament after coming third in the vote – denounced the alliance as 'unholy' and charged the JKPDP with having broken its election promises by joining forces with the BJP.

Efforts to resume India–Pakistan talks over the conflict in Kashmir were given a boost in March, when Subrahmanyam Jaishankar and Aizaz Chaudhry, the Indian and Pakistani foreign secretaries respectively, met in Islamabad. This resulted in the first high-level discussions between the sides since the breakdown of the process the previous August. But, while both sides reiterated their desire to work together, there was no immediate improvement in the bilateral relationship.

Modi and Sharif spoke in July, on the sidelines of the Shanghai Cooperation Organisation summit in Russia, agreeing to organise a meeting on counter-terrorism between their national security advisers. Much to the chagrin of Kashmiri separatist leaders, the two explicitly refrained from discussing Kashmir. Shortly after the meeting, however, high-ranking Pakistani officials stated that any formal bilateral talks would have to address the conflict there. The counter-terrorism summit was eventually cancelled on 23 August, due to the sides' differences on setting the agenda. Islamabad had continued to push for Kashmir to be included in the talks, and Pakistani National Security Adviser Sartaj Aziz had sought to meet with leaders of separatist coalition All Parties Hurriyat Conference while in India – only for the request to be rejected by New Delhi.

On 11 September, a state court in Jammu and Kashmir upheld a colonial-era law banning beef production. The legislation, which had not been enforced for nearly 70 years, punished the slaughter of cows with a fine and up to ten years in prison. The ruling immediately led to protests across predominantly Muslim Kashmir, with the largest demonstration taking place in Srinagar, the summer capital of Jammu and Kashmir. Hardline separatist leader Syed Ali Shah Geelani called for a 'total shutdown' in the state, prompting the authorities to impose curfews in many areas of Srinagar. Yet Geelani changed his position shortly thereafter, asking Muslims to obey the ban lest they exacerbate tensions with local Hindus. Amid widespread unrest in Jammu and Kashmir, the Supreme Court of India imposed a two-month suspension on the beef-production ban. The Supreme Court dismissed on 16 October a petition seeking that the case be heard by three judges in the high court in Srinagar, on the grounds that the move could threaten law and order in the city.

Modi visited Jammu and Kashmir in early November, pledging to boost economic growth in the state by providing it with an aid package worth more than US$12 billion. The trip sparked large-scale protests across Kashmir, leading to the detention of more than 400 alleged separatists prior to the prime minister's arrival, as well as the death of one demonstrator during a clash with police in Srinagar. The aid package was rejected by the Kashmir Chamber of Commerce and Industry, which argued that it failed to adequately address the needs of civilians affected by the devastating floods of the previous year.

Modi and Sharif met again briefly in late November, on the sidelines of the Paris summit on climate change. One week later, it emerged that Pakistan's new national security adviser, Naseer Khan Janjua, had taken part in discussions with his Indian counterpart, Ajit Doval. Their talks, held in Bangkok, reportedly covered counter-terrorism, peace and security, and – surprisingly – Jammu and Kashmir. And bilateral relations appeared to improve further on 25 December, when Modi made an unexpected visit to Pakistan, the first by an Indian prime minister in nearly 12 years. The trip was widely interpreted as reflecting a shift in India's foreign policy towards greater engagement with Pakistan, even as the countries' forces continued to launch artillery strikes across the Line of Control.

Cross-border violence

According to the South Asian Terrorism Portal, Indian and Pakistani forces committed 25 ceasefire violations in the first weeks of January. The incidents included heavy shelling and exchanges of small-arms fire. In the first three months of the year, such clashes were concentrated in the Pura and Samba sectors of the Line of Control, both of which were controlled by India. While commanders from both sides staged in early March a flag meeting at the border, allowing them to reiterate their commitment to upholding the ceasefire, two attacks later that month threatened to destabilise the situation. In the first of these assaults, four militants stormed a police station in Indian-administered Kathua district, killing three police officers and injuring 12 others. The next day, two fighters attacked an army checkpoint in Samba district.

Concern that the conflict would intensify rose sharply in May and June, when at least six former militants and civilian separatists were killed in Sopore, a town in Indian-administered Kashmir. The police blamed a splinter faction of Hizbul Mujahideen for the murders, while the group alleged that the Indian security forces were responsible. Yet no one was convicted of the crimes, sparking fears of a return to the era of unsolved killings seen in Jammu and Kashmir in the 1990s.

There were another eight ceasefire violations across the Line of Control in July, and six more in August. Meanwhile, after Pakistani militants made a series of attempts to cross the border, the Indian security forces announced that they would conduct a general review of their counter-insurgency strategy. According to a Jammu and Kashmir police census published days later, Indian-controlled Kashmir contained an estimated 142 local and foreign militants, most of them affiliated with Hizbul Mujahideen. Another police assessment, published in July, concluded that separatist militancy in the region was increasing for the first time in 15 years. This claim was given further weight on 7 October, when unidentified gunmen killed Altaf Ahmad Dar. A police sub-inspector regarded as one of India's leading counter-terrorism officers in Kashmir, Dar had overseen several operations against Hizbul Mujahideen and Lashkar-e-Taiba.

The same month, Chief Minister Sayeed announced that more than 20,000 concrete bunkers would be constructed along the Line of Control to provide refuge for civilians, 14 of whom had been accidently killed in border clashes since the beginning of the year. While there was a significant decline in such battles during the remainder of the year, former chief minister Farooq Abdullah claimed that the Indian Army was unable to prevent militant incursions, and that only a political solution to the conflict could provide security to civilians in Jammu and Kashmir.

Indian officials' concern about rising militancy was aggravated by the expansion of the Islamic State (also known as ISIS or ISIL) outside the Middle East. By June, the jihadist group's distinctive black and white flag had become a common sight at protests in Srinagar and Rajouri, and was often displayed alongside the flag of Pakistan. Senior figures in the Indian Army stated in July that ISIS sought to establish a presence in Pakistan-administered Kashmir, but later conceded that there was no reliable evidence for the claim. It seemed likely that the demonstrators had used the ISIS flag not to declare their affinity with the group, but rather to irritate the Indian authorities.

Persistent discontent

During the January coalition negotiations, a major point of contention between the BJP and the JKPDP was the Armed Forces Special Powers Act, which granted a range of legal immunities to Indian soldiers operating in Jammu and Kashmir. After the BJP fiercely opposed the JKPDP's demand that the law be revoked, a military spokesperson declared on 18 February that the provisions of the act would remain in place, in light of Pakistan's alleged support of terrorist camps in Kashmir.

Many residents of Jammu and Kashmir remained discontented with the legislation, however, due to widespread accusations of human-rights abuses by Indian security forces. This sentiment was particularly evident in mid-April, when the shooting of a teenager by police prompted hardliners in the All Parties Hurriyat Conference to organise a series of protests in Bulgam district. The state government responded later that month, announcing that it would launch an investigation into 35 alleged human-rights violations by the security forces. In its annual report on human rights worldwide, the US Department of State found that both the security forces and militants in Jammu and Kashmir continued to carry out abuses, especially extrajudicial killings, and that pervasive corruption in the state severely hampered law enforcement. Two human-rights organisations published in September a report that repeated these allegations, suggesting that there was evidence of an 'institutional cover-up' designed to protect dozens of high-ranking military and police officials involved in killings, disappearances, torture and sexual violence in the state. The groups alleged that more than 900 individuals had committed such crimes between 1990 and 2014.

While allowing the security services to retain their far-reaching powers, the authorities also appeared to see developmental efforts as important to ending the conflict. Amid claims in the media that Jammu and Kashmir would soon receive a major development package, the state government announced on 31 July that it would establish a medical college in Anantnag district and appoint more than 800 doctors and 4,000 paramedics. J.P. Nadda, India's health minister, announced shortly afterwards that several medical-science centres would be built in the state. The developmental drive culminated in Modi's announcement of the aid package, which was quickly followed by a federal proposal to create 3,000 state-government jobs in Indian-controlled Kashmir, as well as 6,000 transit facilities designed to aid the rehabilitation of Kashmiri migrants. Adding to the sense that the authorities were addressing public discontent, the Jammu and Kashmir police opened an investigation into paramilitary personnel, in connection with the protester who was killed during the prime minister's visit.

The state government subsequently declared that, from February 2016, it would implement the National Food Security Act, under which food rations would be delivered to all citizens' homes. Yet critics of the scheme alleged that it would prevent many civilians from buying subsidised rations, with separatist leader Mirwaiz Umar Farooq claiming that the quantities of food allocated to each family would lead to starvation. In the wake of these announcements, there was yet another series of demonstrations in Jammu and Kashmir.

Inquiries into human-rights abuses by Indian security forces and large-scale development aid both marked positive developments in 2015. However, civilians continued to be affected by the region's ongoing conflict, particularly during cross-border violence in the summer. As the conflict in Kashmir drags on, human security will remain elusive.

Pakistan*

Key statistics	2014	2015
Conflict intensity:	High	High
Fatalities:	3,500	3,000
New IDPs:	910,000	
New refugees:	290,000	

Throughout 2015, the security situation in Pakistan was shaped by the activities of militant groups and the implementation of the National Action Plan (NAP), the government's

* *The Armed Conflict Survey* classifies Pakistan's militancy in the northwest, insurgency in Balochistan and sectarian violence as a single conflict, thereby reducing the number of active conflicts assessed in the book to 37.

new counter-terrorism strategy. The 20-point plan reinstated the death penalty, established special military courts to try suspected terrorists, and improved oversight of madrassas through licensing and registration schemes. The NAP also built on a series of security operations, including *Zarb-e-Azb* (*Sword of the Prophet*), a military offensive launched in June 2014 that led to the destruction of militant networks and hideouts across the Federally Administered Tribal Areas (FATA). The positive effects of that operation began to be felt in 2015, with an overall decline in large-scale militant attacks. However, the assaults that were carried out disproportionately targeted Shi'ites, who make up around 20% of Pakistan's population. Although the government increasingly focused on combating sectarian violence, the losses incurred by the Shia community amplified criticism of Prime Minister Nawaz Sharif, who was sometimes accused of implementing the NAP slowly and inefficiently. In an effort to address the suffering of civilians, Islamabad focused the majority of its aid efforts on assisting those who were threatened by militant groups, particularly internally displaced persons (IDPs) in the northwest.

Military advances in the northwest

In parallel to *Zarb-e-Azb*, which targeted Tehrik-e-Taliban Pakistan (TTP) and affiliated militant groups, the security services conducted smaller campaigns in Khyber Agency. *Khyber One*, an operation centring on Tirah Valley that was launched in October 2014, was succeeded on 20 March by *Khyber Two*. The Pakistan Army began the combat phase of *Khyber Two* by deploying troops to the Tirah Valley and nearby Sipah; by the end of March, it had captured Mazatal. The following month, the military made a significant advance, removing the TTP from Sipah and Akkakhel.

On 8 May, a helicopter carrying several foreign dignitaries was shot down in Gilgit-Baltistan. The Norwegian and Philippine ambassadors to Pakistan were killed immediately, and the Indonesian ambassador died in hospital a few weeks later. The TTP claimed responsibility for the attack, although it remained unclear whether the group had known that the diplomats were on board. The

TTP carried out on 18 September one of its most high-profile assaults of the year, targeting Pakistani forces at Badaber air base, near Peshawar. There, a team of 14 fighters murdered at least 28 people – 16 of whom were praying at a mosque – before being killed by the security forces.

Two other militant organisations also conducted large-scale bombings in the northwest. In December, Sunni supremacist movement Lashkar-e-Jhangvi killed 24 people at a marketplace in Parachinar, in Kurram Agency; and Taliban splinter group Jamaat-ul-Ahrar killed at least 26 people at a government office in Mardan, in Khyber Pakhtunkhwa.

At a ceremony marking the anniversary of *Zarb-e-Azb*, the Pakistan Army formally ended *Khyber Two*. The military announced that more than 50 Pakistani soldiers had died during *Khyber Two*, which had achieved its stated objective of removing militants from Tirah Valley and Khyber Agency. As a consequence, 95% of Tirah Valley was under government control. Defence Minister Khawaja Asif declared that 2,763 militants had been killed in the campaign, and around 837 of their hideouts destroyed. These figures were difficult to verify, however, due to severe restrictions on reporting from conflict zones in the northwest.

Asif and Prime Minister Sharif announced that *Zarb-e-Azb* would soon begin its final phase: a ground assault in North Waziristan's remote Shawal Valley, thought to be the TTP's last major stronghold in the region. The security forces subsequently made preparations for the undertaking, clearing mountain peaks surrounding the area in July, and conducting in early August a series of heavy airstrikes that reportedly killed at least 65 militants. According to army chief General Raheel Sharif, 203 militants were killed between 18 August, when ground forces entered the valley, and the end of the month.

In September, Pakistan's first indigenously produced unmanned aerial vehicle designed for combat, the *Burraq*, carried out its first combat mission. This made Pakistan the ninth country in the world to field locally produced unmanned aerial vehicles. The mission followed a sharp decline in strikes by US unmanned aerial vehicles, a fixture in northwest Pakistan for nearly a decade.

Independent monitoring groups recorded between ten and 13 such attacks, resulting in between 60 and 85 deaths, in 2015 – the lowest figures for seven years.

By October, the majority of the Pakistan Army's operations were concentrated in Shawal Valley. The military stated that it had carried out several airstrikes in Datta Khel, killing nearly 50 militants, and in Tirah Valley, against the few armed groups that remained there. Asif claimed on 18 October that the campaign in the northwest had led to the killing of all Pakistan-based members of the East Turkestan Islamic Movement, a Uighur separatist group that operated primarily in the Chinese province of Xinjiang, and that had worked with the TTP. The announcement came as Beijing and Islamabad continued to strengthen their economic ties through the China–Pakistan Economic Corridor, an ambitious infrastructure- and energy-investment scheme.

Asif stated in November that 359 army personnel had been killed during operation *Zarb-e-Azb*, and that 89% of North Waziristan and 87% of Khyber Agency had been cleared of militants. Near the end of the month, the Ministry of the Interior stated that militants had caused the deaths of more than 8,500 security personnel and civilians in the preceding five years. Around 1,500 security personnel had been killed in the FATA – more than in any other region – while another 351 had died in Khyber Pakhtunkhwa. According to army spokesman Lieutenant-General Asim Bajwa, at least 3,400 militants and 488 security personnel had been killed during *Zarb-e-Azb* by December.

Furthermore, the security forces faced a new challenge from jihadist group the Islamic State (also known as ISIS or ISIL), affiliates of which claimed responsibility for several small-scale attacks in the northwest, including two assaults in Bajaur Agency. In January, former TTP spokesman Shahidullah Shahid pledged allegiance to the Middle Eastern group, declaring that it had established a caliphate covering Pakistan, Afghanistan and parts of neighbouring countries. The TTP rejected this claim, publishing in May a 60-page document that reaffirmed its long-standing ties to al-Qaeda. In August, social-media accounts linked to ISIS distributed photos and a video purporting to show an ISIS training camp in the FATA. Yet, while the organisation worked to expand its presence in Pakistan and received extensive media coverage there, it remained a relatively minor threat to the country.

Diplomatic crises, domestic controversy

Recognising the links between the conflicts in Pakistan and Afghanistan, Prime Minister Sharif implemented several measures designed to encourage rapprochement with the Afghan government. On 12 May, he travelled to Kabul accompanied by Raheel Sharif and General Rizwan Akhtar, head of Pakistan's Inter-Services Intelligence (ISI), to meet Afghan President Ashraf Ghani. After the sides had discussed bilateral security cooperation for several days, it emerged that the ISI and the National Directorate of Security (NDS), Afghanistan's intelligence service, had signed a draft Memorandum of Understanding on sharing intelligence and improving their coordination in the fight against the Taliban. Yet the move caused an outcry among the Afghan public, and lawmakers in Kabul were quick to point out that the ISI had helped create the Taliban in the 1980s, during the Soviet occupation of Afghanistan. The arrangement was also viewed with suspicion by many Pakistanis, who believed that the TTP had coordinated attacks on their country from Afghan territory.

Largely due to this mistrust, the attempts at rapprochement quickly collapsed. The Afghan government announced on 29 July that Mullah Omar, leader of the Afghan Taliban, had died in 2013 in a Karachi hospital. Following a series of suicide bombings in central Kabul on 7 August, and another outside the city's international airport a few days later, Ghani criticised Pakistan for continuing to harbour militants. A spokesperson for the NDS went so far as to blame the attacks on 'elements within the Pakistani army'. Ghani then publicly called on Islamabad to take decisive action against those in Pakistan who sympathised with the Afghan Taliban, and, shortly thereafter, Afghan Vice-President Abdul Rashid Dostum called for diplomatic relations with Pakistan to be severed.

Prime Minister Sharif was also forced to contend with a diplomatic crisis involving the United States. Islamabad stated in late July that it had secured US$337 million from the US under the Coalition Support Fund (CSF), a programme that contributed US$1 billion per year to Pakistan's anti-militancy efforts in the northwest. But the US Department of Defense threatened on 20 August to withhold the money, due to allegations that the Pakistani government had failed to take action against militant group the Haqqani network. In light of the fact that *Zarb-e-Azb*, *Khyber One* and *Khyber Two* had been paid for largely with CSF funds, Islamabad responded by stating in early September that it would unveil 'compelling evidence that the group had been destroyed'. However, this claim was quickly thrown into doubt by the Haqqani network's release of videos showing one of its training camps, likely located in eastern Afghanistan or North Waziristan. After Islamabad failed to produce evidence that the group had been destroyed, Washington withheld the CSF funds. Nonetheless, the US Department of State confirmed in early October that it had approved US$265m in separate counter-terrorism funding for Pakistan under the Foreign Military Financing (FMF) scheme. Washington also approved in October the sale of eight F-16 fighter jets to the Pakistani military, during a visit to the US by Prime Minister Sharif.

Against the backdrop of rising international tension, the Pakistani government came under growing pressure at home with the dismissal the same month of National Security Adviser Sartaj Aziz, a veteran civilian leader and close confidant of Sharif. Aziz was replaced by career military officer Lieutenant-General Naseer Khan Janjua, whose appointment appeared to signal that the army had strengthened its influence on Pakistani politics. Concerns about this trend had been raised in July by the International Crisis Group, which argued that the implementation of the NAP was allowing the military to entrench its influence on the elected government. And civilian oversight of the security forces was further impaired on 4 November, when the Pakistan Electronic Media Regulatory Authority banned reporting on 72 outlawed organisations, including militant group Lashkar-e-Taiba and its suspected affiliate Jamaat-ud-Dawa.

The media blackout seemed to have been triggered by positive coverage of Jamaat-ud-Dawa's relief operations in the FATA and Khyber Pakhtunkhwa, following an earthquake that killed 226 people there in late October (the Ministry of the Interior claimed in December that Jamaat-ud-Dawa had not been banned outright).

Another source of controversy was the FATA Reforms Bill, presented to the National Assembly in early September. The legislation included a provision overturning Article 247 of the constitution, which stated that neither the Supreme Court nor any high court had jurisdiction in the northwestern territories. The move cleared the way for a debate on merging the FATA with Khyber Pakhtunkhwa. However, the bill was opposed by the Grand Tribal Jirga at a meeting convened in mid-September, who suggested that the FATA be recognised as a province. By the end of the month, a consensus in favour of a merger had begun to emerge among prominent political figures, including the leaders of opposition parties Pakistan Tehreek-e-Insaf and Jamaat-e-Islami.

The debate surrounding the FATA's political future gathered pace in November, when Prime Minister Sharif announced the formation of a committee for implementing reforms in the region. On 12 November, six independent parliamentarians from the region asked the speaker of the National Assembly to delay the 22nd amendment bill, on merging the FATA with Khyber Pakhtunkhwa, to incorporate recommendations by the committee established to discuss such a move. A large rally in support of the merger was held in Islamabad on 17 November, and was attended by several leaders of major political parties, who argued that it would allow civilians living there to access the amenities available to citizens in other areas of Pakistan. Tribal leaders demanded on 18 November that all stakeholders and institutions nationwide support the proposed constitutional reforms.

Combating sectarian violence

At least 136 people were killed in sectarian attacks in Pakistan during the first three months of 2015. The year began with two major assaults on the Shia community. Eight Shi'ites died in a bombing outside a congregation hall in Rawalpindi on 9 January. Three weeks later, 60 people were killed by a suicide bombing at a Shia mosque in Shikapur, the deadliest sectarian attack since 2013. The security forces initially blamed Lashkar-e-Jhangvi for the operation, although it later emerged that responsibility lay with TTP splinter group Jundallah. Pakistan's non-Muslim minorities were also targeted. On 15 March, coordinated suicide attacks on two churches in Lahore caused the deaths of 15 people, including two policemen.

In a review of the NAP released in February, the government claimed that the security forces had made 416 sectarianism-related arrests and recorded 547 incidents of hate speech since the implementation of the strategy, which included several measures designed to curb sectarian tensions. Interior Minister Chaudhry Nisar Ali Khan optimistically predicted that sectarian violence would quickly decline, as the security forces stepped up their efforts to implement the NAP. Shortly thereafter, the Milli Yakjehti Council, an alliance of senior Muslim clerics and major religious parties, offered their support to government efforts. Islamabad moved to strengthen its links with other religious bodies, establishing in July several committees to promote interfaith dialogue, including the National Committee on Sectarian Harmony. The government also pressured existing religious organisations, such as the Pakistan Ulema Council, to promote a message of tolerance. As a consequence, the council issued in August a code of conduct for new students in madrassas, and pledged its 'unconditional' support to the NAP.

The Sharif administration put forward in April the Prevention of Electronic Crimes Act 2015, which criminalised anyone who 'prepares or disseminates' electronic communications that praise a person 'accused of a crime' or 'advance religious, ethnic or sectarian hatred', and which defined the 'glorification' of crimes online as 'cyber-terrorism'. Reportedly written with minimal input from legal and technological experts, the bill raised domestic and international concerns about privacy and freedom of expression, and was not passed in 2015.

Prime Minister Sharif announced on 13 May a series of initiatives to combat sectarian violence, following an attack by Jundallah that killed 46 Ismaili Shi'ites travelling aboard a bus in Karachi. Although the group claimed responsibility for the assault on behalf of ISIS, this was disputed by the authorities. The government maintained that ISIS had no discernible presence in Pakistan, and suggested that Lashkar-e-Jhangvi had carried out the attack to access the Middle Eastern group's extensive financing network. Meanwhile, the security forces launched a large-scale counter-terrorism operation, arresting hundreds of people in connection with the attack, and Islamabad announced that it would improve policing and surveillance in Karachi's suburbs, and would conduct weekly apex committee meetings between political and military leaders to discuss the NAP. However, the government appeared to make little progress in implementing these new measures.

Shortly after the killings in Karachi, Pakistan's anti-terrorism court made use of the NAP to indict 106 people in connection with the November 2014 lynching of a Christian couple in Punjab. The authorities also implemented the part of the strategy that called for monitoring Pakistan's 13,000 madrassas, announcing in June a plan to target those suspected of promoting sectarianism.

In August, Punjab Home Minister Shuja Khanzada was killed in a suicide bombing carried out by Lashkar-e-Jhangvi, whose leader, Malik Ishaq, had died during a shoot-out with police officers the previous month. Ishaq had been gunned down along with his two sons, another Lashkar-e-Jhangvi commander and ten other militants. His death followed the execution in January of Akram Lahori, the group's co-founder, who had been convicted of involvement in more than 30 sectarian murders. Khanzada had been a staunch opponent of sectarianism and, in the wake of his death, Islamabad announced the creation of a committee to oversee a crackdown on militancy and other crime in Karachi. Senior government figures rushed to express their desire to combat sectarian violence, with Interior Minister Khan announcing on 7 September that the next phase of implementing the NAP would focus on the issue. A few days later, the government unveiled legislation that made it a crime to describe non-Sunnis as 'infidels'.

Lashkar-e-Jhangvi incurred another loss in late November, when co-founder Haroon Bhatti was killed in similar circumstances to Ishaq. Bhatti had been extradited to Pakistan from Dubai in October, and was reportedly working as a police informer. Like Ishaq, Bhatti was leading the police to a suspected militant safe house in Islamabad, when militants attacked their convoy. The deaths of Bhatti and Ishaq led to speculation that the shoot-outs had been orchestrated by the security forces, which had found a convenient way to dispose of the militant leaders.

Sectarianism was a particularly prominent issue in Balochistan. Long targeted by militant groups, the Hazara Shia minority in Quetta suffered a number of attacks in June. In one incident, gunmen killed five Hazara shopkeepers in the Bacha Khan Chowk area, triggering demonstrations against the government's failure to end such violence. Human-rights activists and journalists were also attacked in the province. In April, prominent activist Sabeen Mahmud was shot dead in Karachi, shortly after hosting a seminar on kidnappings and disappearances in Balochistan. Zafarullah Jatak, a Baluch journalist, was killed in his home in Quetta in June. Officials attributed both of the deaths to militants, but there were lingering suspicions among activist groups as to whether the killings were carried out by militants or Pakistani security forces. On 14 October, during the Shia holy month of Muharram, Interior Minister Khan announced the deployment of more than 16,000 security personnel across Pakistan to prevent sectarian attacks. Yet, despite this measure, Lashkar-e-Jhangvi conducted two major suicide bombings targeting Shi'ites. The group killed ten people at a mosque in Balochistan's Kachhi district on 22 October, and, one day later, 22 others at an Ashura celebration in Jacobabad, in Sindh.

Efforts to combat such violence were at the centre of a political row that began in Sindh on 6 December, with the expiration of special policing powers granted to paramilitary the Pakistan Rangers. Having held an extended session on the issue, the province's parliament failed to reach a deal on extending the mandate of the force, long a source of tension within the ruling Pakistan People's Party. There were concerns in the Pakistani media that Islamabad would intervene directly in the dispute to allow the force to continue its anti-militancy campaign in Karachi – and these proved well founded on 23 December, when the Ministry of the Interior granted a 60-day extension to its policing powers.

The war in Syria cast an increasingly long shadow over sectarianism in Pakistan. There were reports in mid-December that Tehran was recruiting Pakistani Shi'ites to Zeinabiyoun, a unit that fought in the conflict on the side of the Syrian regime. According to some estimates, hundreds of Pakistanis had joined the war. The reports on the recruitment programme created concern about a backlash against Shi'ites in Pakistan, as both Tehran and Syrian President Bashar al-Assad were loathed by hardline Sunni groups. The fear appeared to have been realised later that month, when Lashkar-e-Jhangvi carried out a suicide bombing at a marketplace in Parachinar, and stated that the attack was intended to warn Pakistani Shi'ites to stay out of Syria.

Despite these developments, two unusually optimistic reports on sectarian violence were published late in the year. The Center for Research and Security Studies released figures indicating that sectarian militancy had declined in the last three months of 2015, although it noted that around 1,200 civilians and security personnel had been killed in that period. Meanwhile, the Human Rights Commission of Pakistan released a study claiming that there had been 69 sectarian killings in the first nine months of 2015, significantly fewer than the 97 such murders reported in the same period in 2014.

Pakistan executed more than 300 people in the year, a greater number than either the US, Iraq or Saudi Arabia, as Pakistan began to again carry out executions following the end of the moratorium on the death penalty. The Pakistani Supreme Court had ruled in August that secret military courts were legal, and could sentence civilians to death. On 29 August, the authorities executed 18 militants, some of whom had been convicted of sectarian killings. Four days later, a military court sentenced five people to death over their involvement in sectarian murders and an attack on a school for girls. Nonetheless, Amnesty International reported that the majority of those who were executed had not been convicted of terrorist offences. By the end of the year, between 6,000 and 8,000 Pakistanis remained on death row.

Rising insurgency in Balochistan

As well as a growing number of sectarian attacks, Balochistan's authorities were forced to contend with an insurgency that intensified throughout the year. The province's separatist groups continued to carry out gun assaults, bombings and targeted killings (reporting on such incidents was affected by the Pakistan Electronic Media Regulatory Authority's ban on media coverage of militant activity). Spurred by a series of agreements linked to the China–Pakistan Economic Corridor, Islamabad emphasised its willingness to find a political solution to the conflict.

In January, new divisions emerged between the members of Balochistan's shaky governing coalition: the National Party, the Pakistan Muslim League–Nawaz (PML–N) and the Pakhtunkhwa Milli Awami Party (PMAP). All three claimed an exclusive right to choose the next mayor of Quetta, and they only reached a deal on the issue on 27 January, one day before the mayoral elections, when the PML–N and the PMAP agreed to endorse Kaleemullah Kakar – who went on to win the vote.

During a visit to Pakistan in April, Chinese President Xi Jinping joined Prime Minister Sharif in unveiling plans for the China–Pakistan Economic Corridor. Under the scheme, energy and infrastructure projects costing a total of US$46bn would link China's western Xinjiang region with the port of Gwadar, in southern Balochistan. The announcement was enthusiastically endorsed by many Pakistani leaders, with Prime Minister Sharif stating that the corridor would transform Pakistan into a regional economic hub. However, concerns were immediately raised about the feasibility of the undertaking. Several Baluch separatist leaders, including the Baloch Republican Party's Brahamdagh Bugti, voiced their opposition to the plan, arguing that Islamabad was allowing Beijing to invest in occupied territory. Balochistan's centrality to the corridor was also controversial due to fears about the insurgency: it seemed likely that that the new projects would be targeted by the myriad armed groups operating in the province.

The Pakistani government maintained that it was close to finding a political solution to the insurgency, with Abdul Qadir Baloch, minister for states and frontier regions, suggesting in late March that the government sought to achieve this through negotiations. Accordingly, in mid-July, Balochistan's provincial government sent a delegation of influential tribal leaders to enlist the help of Suleman Dawood, the Khan of Kalat, who had significant influence among Baluch separatist leaders. Sanaullah Zehri, leader of the Balochistan faction of the PML–N, announced on 21 July that he would also hold talks with Dawood. Abdul Malik Baloch, the province's chief minister, stated that both his government and Islamabad intended to negotiate with the separatists.

It was reported in August that nearly all major Baluch insurgent groups were engaged in negotiations with the Pakistani government. Sharif approved on 6 August the Peaceful Balochistan Plan, which had the declared aim of bringing the Baluch people 'back into the political mainstream'. The initiative pledged job training and a payment of between US$5,000 and $15,000 to each militant who surrendered. These renewed efforts to end the conflict were widely viewed as resulting from the agreement on the China–Pakistan Economic Corridor, particularly after Prime Minister Sharif announced in late August that 11 of the project's 33 economic zones would be located in Balochistan.

The provincial government enacted in December the 2013 Murree Agreement, which stipulated that Chief Minister Baloch would resign half-way through his five-year tenure, and would be replaced by Zehri. Although there had been doubts that the deal would be implemented in a timely fashion, the transfer of power went ahead at the end of the month.

Working in line with the NAP, Balochistan's authorities claimed to have made more than 2,000 arrests of individuals suspected of militant activity in the first three months of 2015. Yet, despite this, attacks by armed groups increased in both intensity and frequency during that period. Between April and June, operations were conducted by all major separatist insurgent groups, including the Baloch Liberation Front (BLF), the Baloch Republican Army, the Baloch Liberation Army (BLA) and the United Baloch Army (UBA). Several particularly fierce clashes took place in mid-May, when the police carried out raids in Kalat, Mastung and Boi districts.

While Balochistan's separatists primarily targeted security personnel, they also launched several major assaults on civilians. On 10 April, BLF fighters attacked a camp near Turbat, killing 20 construction workers. Allah Nazar, the group's leader, justified the assault on the grounds that the victims had been employed by the Frontier Works Organization, a firm linked to the Pakistani military. Another 22 civilians, most of them Pashtuns, were killed on 29 May in Mastung. There, UBA gunmen disguised as members of the Pakistani security forces executed the civilians after ordering them off the buses they were travelling on (drawing comparisons with the attack on Shi'ites in Karachi earlier that month). A massive security operation was launched in Balochistan in response to the murders, with paramilitary the Frontier Corps targeting sectarian militias, smugglers and criminal groups across the province.

There were four major assaults targeting residents of Balochistan in September and October. Ten people were killed in a bombing at a bus station in Multan on 13 September, an incident that the Pakistani authorities blamed on the BLA. On 19 October, an explosion aboard a bus in Quetta left 11 people dead, though no group claimed responsibility. And the sectarian attacks in Kachhi and Jacobabad took place shortly thereafter. The first two operations occurred despite the government's efforts to reach out to separatist leaders, and suggested that militants in the province would continue to target civilians.

There was also an outbreak of violence near Balochistan's border with Iran, as Panjgur, Mand, Washuk and Chagai were periodically caught in Iranian forces' artillery strikes on militants. After a spate of attacks along the frontier in April, Balochistan's government attempted to strengthen its border-security and counter-terrorism cooperation with Tehran. This culminated in a meeting between Secretary-General Seyfollah Cheta and Iranian officials on 29 April, at which both sides pledged to devote more resources to these efforts.

Following the violence in April and May, Prime Minister Sharif met with provincial leaders in Balochistan and Lieutenant-General Janjua, then head of the Pakistan Army's Southern Command, to formulate a stronger response to militancy in the province. This was complemented by a drive to persuade militants to surrender – an undertaking that met with some success in August, when more than 400 militants from various insurgent groups gave themselves up at a ceremony in Quetta. Several small groups of militants also surrendered to security forces in other locations across Balochistan.

The same month, the Pakistan Army announced that it had killed Nazar. While government officials, including Balochistan Home Minister Sarfaraz Ahmed Bugti, were quick to confirm the militant leader's death, this proved premature: Nazar appeared in a video in late November to state his opposition to planned Chinese investment in the region. This exacerbated fears that militants would disrupt the expensive projects undertaken as part of the China–Pakistan Economic Corridor. Such concerns had led in August to the formation of the Gwadar Safe City Project, a 1,100-strong force designed to bolster the military presence there. Nonetheless, the army attempted to play down such fears. Lieutenant-General Amir Riaz, who replaced Janjua as head of the Southern Command, stated in November that there was 'no threat of insurgency in Balochistan'.

Multiplying threats to civilians

As well as attacks by militants, civilians in Balochistan were allegedly subjected to frequent human-rights abuses by the Pakistani security forces, particularly in the wake of large-scale security operations. Reports of such incidents came primarily from Baluch-rights organisations, and were difficult to verify due to the severe restrictions placed on the media.

The scarcity of water in the province also became a prominent issue in 2015. On 21 February, Chief Minister Baloch petitioned the federal government to provide financial and technical assistance for addressing Balochistan's water shortage. He also met with a senior official from the World Bank in early April, repeating his earlier warnings that the province faced acute water shortages due to climate change. Indeed, experts predicted that, if no action was taken, the rapid depletion of underground water would cause a drought in the province in 10–12 years. Quetta and Gwadar were likely to suffer the most severe effects.

Meanwhile, the status of Afghan refugees in Balochistan continued to be deeply controversial. Islamabad announced its intention to repatriate 1.5m Afghan refugees in the country over two years, after refusing to extend their stay past the 31 December 2015 deadline. The government also attempted to address the problems facing civilians displaced from the FATA, announcing in March that more than 200 families would be returned to South Waziristan, and in April that thousands more would be returned to North Waziristan. They were allowed to go back on the condition that they pledged allegiance to the government, and promised not to aid militants, nor to form local militias. Many welcomed the chance to return, but some tribal leaders were angered by Islamabad's demand that they be held collectively responsible for preventing militants from establishing a presence in their territory. The government implemented in July the second phase of its programme for aiding IDPs, providing more than 100 families displaced by fighting with transportation home and a living allowance. Beijing also pledged US$10m to the programme, while the International Organization for Migration stated that it would provide shelter and household items to civilians who had been displaced.

Although Islamabad accepted this assistance, it was overtly hostile to some entities providing aid. In June, the government accused Save the Children of 'anti-state activities', a claim that led to the forced closure of the charity's offices across Pakistan. Save the Children had been under intense scrutiny in Khyber Pakhtunkhwa since 2011, due to suspicions among locals that its vaccination campaign was a front for CIA operations. However, the charity's head office in the country was quickly reopened, and it was given a conditional six-month permit to work everywhere except for the FATA, Balochistan and Gilgit-Baltistan.

Overall, the security services made significant progress in disrupting militant networks across the country. Yet, as military operations in the northwest wound down, Islamabad faced major challenges in shaping the political future of the FATA. The NAP was likely to continue improving security in the near future, but there remained doubts about its ability to curb militancy in the long term.

Chapter Six

Asia-Pacific

China (Xinjiang)

Key statistics	2014	2015
Conflict intensity:	Low	Low
Fatalities:	400	200
New IDPs:		
New refugees:		

Despite undergoing only limited changes on the ground, the conflict in Xinjiang led to far fewer recorded fatalities in 2015 than in the previous year. This apparent shift reflected either improvements in the region's security or a tightening of restrictions on journalists. Violence in Xinjiang continued to follow well-established patterns: Uighurs carried out knife attacks before being shot by police, or else were killed by the authorities in battles and unexplained incidents. There was little discernible strategy behind the terrorist assaults, which seemed motivated by a desire to resist the state by creating disruption and insecurity. The attacks most often targeted security personnel or local officials, and the majority took place in Xinjiang's southwestern prefectures of Aksu, Hotan and Kashgar. Zhang Chunxian, head of the Xinjiang Communist Party, said on 4 November that 'south Xinjiang is the central battleground for fighting terrorism and threats to stability', promising to encourage 'exchanges and blending' between ethnic groups and to expand programmes for teaching Mandarin. According to data

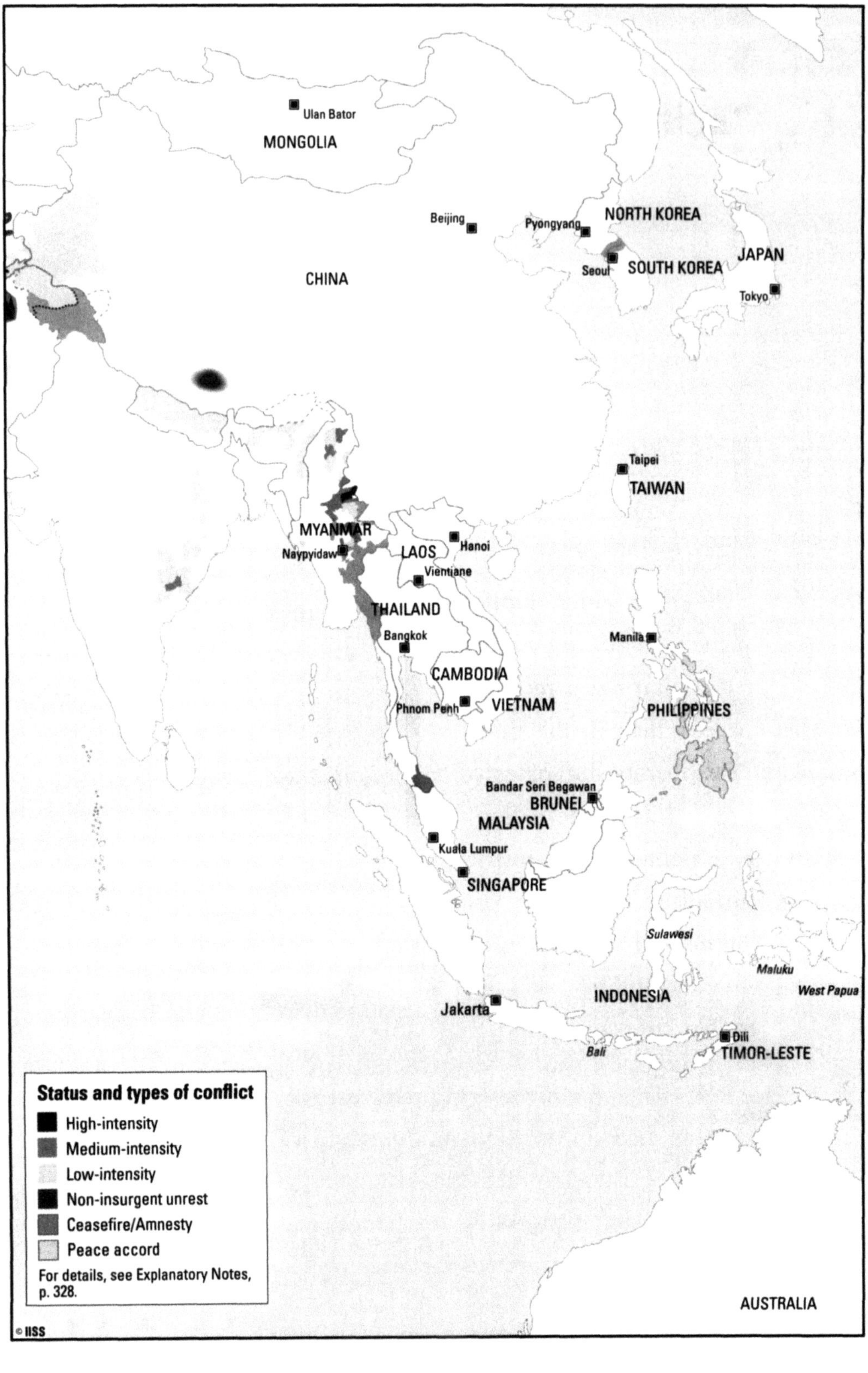
Ulan Bator
MONGOLIA
Beijing
Pyongyang
NORTH KOREA
Seoul
SOUTH KOREA
JAPAN
Tokyo
CHINA
Taipei
TAIWAN
MYANMAR
Naypyidaw
LAOS
Hanoi
Vientiane
THAILAND
Bangkok
Manila
CAMBODIA
Phnom Penh
VIETNAM
PHILIPPINES
Bandar Seri Begawan
BRUNEI
MALAYSIA
Kuala Lumpur
SINGAPORE
Sulawesi
Maluku
West Papua
INDONESIA
Jakarta
Dili
Bali
TIMOR-LESTE
Status and types of conflict
High-intensity
Medium-intensity
Low-intensity
Non-insurgent unrest
Ceasefire/Amnesty
Peace accord
For details, see Explanatory Notes, p. 328.
AUSTRALIA
© IISS

published by the Chinese government in May, 181 'terror groups' in Xinjiang had been destroyed in the preceding year, following the launch of a 'strike hard' campaign in response to several major attacks.

The first reported violent incident of 2015 occurred on 12 January, when, according to state media outlet Xinhua, six people carrying explosives were killed by police officers in Kashgar Yengisheher County, in Kashgar. This was followed in February by an attack in Aksu in which nine suspected terrorists, four police officers and four civilians were killed, and in March by an assault in Kashgar city that led to the deaths of four militants. Three police officers and three suspected terrorists were killed between 11 and 12 May, during successive attacks on a police checkpoint in Hotan's Luopu County. By 13 May, the police had arrested more than 200 people in connection with the attacks. On 22 June, at least three police officers and 16 suspected terrorists were killed in a knife and bomb attack in Tahtakoruk district, on the outskirts of Kashgar city.

One of the deadliest attacks ever reported in the region took place on 18 September, when Uighur men killed at least 50 people and wounded around 50 others, most of them ethnic Hans, at Sogan coal mine in Baicheng County, in Aksu. The militants used knives to assault the site's security personnel and then miners in their dormitories, in what the government called a 'long-planned, well-prepared, large-scale attack by separatists against police officers and mine owners'. An official report stated that the operation had been planned by two men radicalised by extremist videos, one of whom had met with foreign groups. But the head of a local farming unit earlier suggested that the attackers had been radicalised by policies of forced cultural assimilation, such as those preventing women from wearing veils. It appeared that the families of the perpetrators had been repeatedly humiliated by state personnel seeking to enforce these policies. As in previous assaults, the militants targeted economic infrastructure owned by ethnic Hans. The Ministry of Public Security announced on 20 November that the security forces had killed 28 people in pursuit operations connected with the attack, while Radio Free Asia reported that 11 of the dead were women and children.

Security incidents involving Uighurs outside Xinjiang appeared to occur more often than in previous years. The most destructive attack took place on 6 March in Guangzhou, in Guangdong Province, where three men with knives assaulted a crowd at a railway station, injuring 13 people. During the incident, which was reminiscent of attacks on railway stations in 2014, the police shot dead one assailant and arrested another. Security operations to arrest Uighurs suspected of terrorism were conducted across China, leading to the deaths of at least five people in Shenyang, in the northeastern province of Liaoning, and Zhengzhou, the capital of Henan Province. On 7 December, four people suspected of planning terrorist attacks were arrested in Taiyuan, in Shanxi Province.

Transnational militancy

While Beijing had long pointed to the influence of outside forces in Xinjiang, security and human rights in the region took on an increasingly international dimension in 2015. This was partly due to purported links between Uighur militants and jihadist group the Islamic State (also known as ISIS or ISIL), which released in December a recording of a chant in Mandarin urging Muslims to 'take up weapons and fight', having earlier executed a Chinese hostage. Zhang said in March that the authorities had arrested people who had joined ISIS and returned to China 'to participate in terror plots'. In June, he stated that there had been 'escalated infiltration and sabotage activities from foreign hostile forces that are disguised as religion'.

There was some evidence of ties between Uighur militants and transnational jihadists: two Chinese Uighurs alleged to be members of ISIS were arrested while attempting to enter Turkey from Syria in April. Furthermore, the Turkestan Islamic Party, an organisation comprising Uighurs from Xinjiang, was known to be fighting alongside al-Qaeda affiliate Jabhat al-Nusra in Syria.

Violent incidents in Southeast Asia also spoke to the involvement of Uighurs in transnational militant networks. Uighur refugees often travelled through the region on their way to Turkey (where there is popular sympathy for the plight of fellow ethnic Turkic peoples). The 17 August bombing of the Erawan Shrine in Bangkok – which killed 20 people, most of them ethnic Chinese – appeared to have been carried out by a network of Uighurs and Turkish nationalists. Several Uighurs were arrested in Thailand and Malaysia in connection with the attack, and key suspects were traced to Turkey. The incident followed Thailand's repatriation of 109 Uighur Chinese citizens on 9 July, for which the country was strongly criticised by the international community. China's Ministry of Public Security claimed that those who had been repatriated were travelling through Thailand to join ISIS, after being recruited by gangs and 'stirred up' by separatist group the East Turkestan Islamic Movement (ETIM) or the World Uyghur Congress, an activist organisation based in Germany.

In Indonesia, three Uighurs were sentenced in July to six years in prison for attempting to join a terrorist group in Sulawesi. Another Uighur, described by the authorities as a potential suicide bomber, was arrested in December, along with eight other people, on suspicion of planning an attack in Jakarta.

Xinjiang's strategic importance

The Xinjiang Uighur Autonomous Region marked its 60th anniversary on 1 October. Yu Zhengsheng, chairman of the National Committee of the Chinese People's Political Consultative Conference, said in a speech to mark the occasion that long-term stability and security were the government's top priorities in the region, and that 'counterterrorism is the focus of our current work'. Xinjiang's location meant that it played a key role in China's Silk Road Economic Belt, an international initiative for which the country established in February a US$40 billion infrastructure fund. Given such large-scale investment in Xinjiang's connectivity with neighbouring countries, Beijing was unlikely to change its security policy in the autonomous region.

President Xi Jinping visited Pakistan on 20 April, signing an agreement for investing US$46bn in the construction of the China–Pakistan Economic Corridor (CPEC), a network of roads, railways and energy pipelines that would run from Kashgar to the Pakistani port of Gwadar. Pakistani Prime Minister Nawaz Sharif told Xi at the time that 'Pakistan considers China's security as important as its own security'.

Beijing continued to strengthen security cooperation with its neighbours. A spokesman for the Chinese People's Political Consultative Conference said on 2 March that an 'anti-terrorism cooperation mechanism', mainly aimed at the ETIM, had been established with more than ten partner countries to exchange information. Beijing had also become a key player in the Afghan peace process, perceiving threats to China's security in the jihadism and instability prevalent in Afghanistan and Pakistan, where Uighur militants were known to have established a presence. As a consequence, China reportedly leant on Pakistan to support talks between Kabul and the Afghan Taliban. During a visit to Beijing on 12 August, Pakistani President Mamnoon Hussain declared that 'almost all' Uighur separatists in Pakistan had been 'eliminated' in military operations.

Beijing also passed on 1 July a new law on national security. The legislation was written in broad terms, providing room for interpretation by the authorities, and did not explicitly mention Xinjiang. Nonetheless, Article 28 speaks of increasing capacity to handle terrorism and extremism, by 'strictly punishing violent terrorist activities'. Beijing adopted on 27 December its first terrorism bill, having theretofore dealt with terrorism cases under criminal law. The new legislation defined terrorism as 'any proposition or activity that, by means of violence, sabotage or threat, generates social panic, undermines public security, infringes personal and property rights, and menaces government organs and international organisations with the aim to realise certain political and ideological purposes'. Scheduled to take effect from 1 January 2016, the law included a clause banning the dissemination of false information on terrorist attacks, and of details about actual incidents. The new legislation also enabled the government to freeze the accounts of people suspected of financing terrorism.

The authorities also displayed a tougher attitude towards dissent among their own ranks. Zhao Xinyu, former editor of *Xinjiang Daily*, was expelled from the Communist Party on 2 November for 'public opposition' to the party line on separatism, extremism, religion and other issues, and for 'improperly' discussing policies. Later that month, Xu Hairong, head of Xinjiang's Commission for Discipline Inspection, announced that some Communist Party members in the region had been 'swayed' by 'pan-Islamism and pan-Turkism', and had participated in terrorist activities. Minister of Public Security Guo Shengkun declared that state officials who failed to satisfactorily implement counter-terrorism measures would be held accountable 'in accordance with law and Party discipline'.

Counterproductive security measures

Civilians in Xinjiang faced an array of threats: militant attacks on ethnic Hans, security operations that were often heavy-handed and the state's repression of local culture. State media reported on 22 January that more than 27,000 suspected criminals were arrested in the region in 2014, an annual increase of 95%, following the launch of the 'strike hard' security campaign, which began in May 2014 and continued into 2015. The *Dui Hua Human Rights Journal* reported in March 2015 that nearly 30,000 criminal trials had been concluded in Xinjiang in 2014, an annual rise of 40%. The publication argued that this trend indicated the government was increasingly suppressing dissent and activism in Xinjiang. There were also reports that the authorities used forced labour to ensure stability.

According to a government White Paper on human rights published on 8 June, ethnic minorities' freedom of religious belief had been 'fully guaranteed'. However, Xinjiang's authorities continued their clampdown on the observation of Islamic customs, an approach that was widely seen as one of the main drivers of the conflict. On 29 April, the local authorities in Aktash, a Muslim-majority village in Hotan, ordered shop owners to sell alcohol and cigarettes or face closure. A local Communist Party secretary was quoted as saying that 'we have a campaign to weaken religion here and this is part of that campaign'.

Similar efforts were made before and during Ramadan (which began on 18 June). Xinjiang's authorities put the security forces on high alert, and implemented measures to prevent Muslims from observing the traditional fast. The US Department of State's latest International Religious Freedom Report stated that China's Uighur Muslims 'experienced severe societal discrimination', and that Beijing 'sought the forcible return of ethnic Uighurs' who had left China. The report also highlighted restrictions on clothing, education and religious practices in Xinjiang. On 10 January 2015, the region's legislature banned the wearing of burkas in public places in Urumqi, mirroring similar measures imposed the previous year, which applied only in some locations or at certain times. In November, the Supreme People's Court criminalised forcing others to wear unspecified clothing or symbols associated with 'extremism'.

Overall, the conflict in Xinjiang appeared to diminish in 2015. Viewing poverty as the underlying cause of the instability there, Beijing remained committed to a combination of economic development and assertive security policy in the region. Nonetheless, there were indications that the government had increased its restrictions on Uighur cultural practices, potentially exacerbating locals' discontent with the state.

Myanmar

Key statistics	2014	2015
Conflict intensity:	Medium	Medium
Fatalities:	300	600
New IDPs:	6,000	
New refugees:	16,000	

Myanmar's civil conflict appeared to move closer to ending on 15 October 2015, when a nationwide ceasefire agreement (NCA) was signed in Naypyidaw following years of negotiations. However, seven of the 15 ethnic armed groups invited to participate in the deal declined to do so, and several other militant organisations had been excluded from the arrangement all along. Among the signatories, only the Karen National Union (KNU) and the Shan State Army–South (SSA–S) had a significant number of fighters. Furthermore, the draft NCA was signed after the parties pushed back contentious issues to the next stage of the peace process, casting doubt on the credibility and durability of the agreement. The deal also seemed to have divided the ethnic armed groups along geographical lines, with those based in the north refusing to sign.

The peace process had nearly been derailed in February, after the Myanmar National Democratic Alliance Army (MNDAA) unexpectedly launched an offensive in northeastern Shan State. This prompted the government to insist that the group and two of its allies be prevented from singing the NCA. The ethnic armed groups had long said that they would refuse to sign if any of them were excluded – yet only five of the groups maintained this position following the offensive. The landslide victory won by Aung San Suu Kyi's National League for Democracy (NLD) in elections held on 8 November renewed hope that the peace process would succeed, but also complicated the negotiations.

Pending transfer of power

All groups that participated in the talks, including those who did not sign the NCA, agreed on the text of the deal and fundamentally wanted peace, as did the government. The agreement provided for, inter alia, joint monitoring, an end to offensive troop movements, demining programmes and the coordination of emergency relief. Prior to the signing of the NCA, the government removed the eight compliant groups from its list of unlawful associations – a step towards their integration into Myanmar's political and economic system. On 18 November, the signatories and the Tatmadaw agreed on a code of conduct.

Although it was possible that the seven groups who refused to sign the deal would do so later, this was unlikely to happen until the political scene settled. Having won the November vote, the NLD was scheduled to form a government in March 2016. The party's supermajority ensured that it would also choose the next president. The elections occurred peacefully, and the limited voting fraud that occurred had little impact on the results. Both the Tatmadaw and President Thein Sein released statements congratulating the NLD on its victory, while the ethnic armed groups signalled their approval of the result.

An NLD spokesperson subsequently stated that Suu Kyi would lead the peace process following the transfer of power, adding that the structure of the talks and key stakeholders might be changed. Suu Kyi said in December that she agreed with the text of the NCA, but had refrained from signing it because some ethnic armed groups had been excluded.

The drafting of a framework for the Union Peace Conference, the political dialogue that forms the next stage of the peace process, began on 1 December and was approved by the Union Peace Dialogue Joint Committee two weeks later. The conference was to have around 700 participants, and the ethnic armed groups who refused to sign were to act as observers. On 19 December, the president's office announced that the Union Peace Conference would begin on 12 January 2016.

Intensifying violence

The government had announced in 2009 that the MNDAA had been defeated. Yet, on 9 February 2015, the group and its allies, the Arakan Army and the Ta'ang National Liberation Army (TNLA), launched a surprise offensive in Shan State's Kokang region. The MNDAA's exiled leader, Peng Jiasheng, appeared to have ordered the campaign to seize control of Kokang and thereby force his way into the peace process. The Kachin Independence Army (KIA) was also involved in heavy fighting with government troops around the same time, but it was unclear whether these operations in neighbouring regions were connected with the fighting in Kokang.

The Tatmadaw, usually reluctant to discuss casualties, began to regularly publish fatality statistics on the battle in Kokang, perhaps to garner public support for its efforts, and to show that it was fulfilling its duties to preserve national unity and protect Myanmar's territorial integrity. The army stated that 47 soldiers, seven police officers and two rebels had been killed in the first week of fighting; the rebels, who had secured positions on high ground, reported that more than 100 soldiers had been killed in the period. Having incurred heavy losses, government forces deployed assault helicopters, fighter jets and possibly unmanned aerial vehicles.

On 17 February, the government declared a 90-day state of emergency and martial law in Kokang, granting the military administrative and judicial powers there. Naypyidaw repeatedly rejected the MNDAA's requests for talks. The clashes raised tensions with Beijing due to the large numbers of refugees that fled into China, and the diplomatic situation grew worse on 13 March, when the Myanmar Air Force dropped bombs on the Chinese side of the border, killing five Chinese citizens. Civilians were displaced on a scale unseen for several years: the Red Cross Society of China said on 19 February that 100,000 people had entered the country to escape the fighting. Beijing announced on 2 March that it had provided temporary settlements for around 14,000 Myanmar citizens in Yunnan, and on 6 March that 4,000 displaced persons had returned to Kokang. Lieutenant-General Wai Lwin, then Myanmar's defence minister, announced on 17 May that there were more than 88,000 Kokang refugees in nine camps in China. Near the end of the month, the Tatmadaw stated that Kokang was under its 'total control', and that its soldiers occupied all strategically important hilltops in the region. It was estimated that more people died in the battle for Kokang than in the entire conflict in 2014.

The MNDAA declared a unilateral ceasefire on 10 June, after which fighting in the region gradually subsided. The TNLA, the MNDAA and the Arakan Army issued on 5 August a joint statement calling on the government to negotiate with them on bilateral peace agreements. But Senior General Min Aung Hlaing, head of the Tatmadaw, responded weeks later that there was 'no prospect for peace' with the MNDAA, and that both the group and the KIA's 4th Brigade, based in northern Shan State, would have to surrender their weapons. The fighting briefly flared up again in late September, but calmed down enough for the government to revoke the declaration of martial law in Kokang in November.

Outside the region, the Tatmadaw often engaged in battles with the KIA and the TNLA, reportedly leading to the deaths of dozens of people in Shan and Kachin states. The clashes may have been linked with China and Myanmar's attempt in January to tighten border security along known smuggling routes, affecting the illicit hardwood exports that were an important source of income for many Myanmar armed groups. In Kachin State, the fighting centred on the jade-mining town of Hpakant. On 15 January, the KIA abducted the state's transport minister and three police officers travelling in his convoy, leading to three weeks of violence. In late March, there was large-scale fighting in Mansi township, a centre for illegal logging, where more than 20 people were reported to have been killed, and the government deployed airpower to fight the rebels.

In an effort to dislodge the KIA, the Tatmadaw launched a renewed air and ground offensive in Hpakant, as well as against one of the group's bases in Mali Yang, in Sumprabum township. By 22 July, more than 1,000 civilians in the areas had been forced to flee, and a similar number had been displaced in Kayin State, following brief clashes between the military and the Democratic Karen Benevolent Army. In late July, the army battled the Shan State Army–North (SSA–N) near its headquarters in the ruby-mining township of Mong Hsu. On 21 September, during another outbreak of fighting in Mansi, the Tatmadaw reportedly captured two rebel outposts.

There were fears that the government and the armed forces had exploited rivalries between the ethnic armed groups to justify a strengthened offensive against those who had refused to sign the NCA. This was suggested by the fact that nearly all major southeastern groups agreed to sign the deal, but the Tatmadaw intensified its operations against northern rebels, despite the government's promise that it would not attack non-signatories. The SSA–S suffered several attacks by the armed forces as it deliberated whether to sign the NCA.

In what was perhaps an attempt to pressure the SSA–N into signing the ceasefire agreement, the Tatmadaw also launched an offensive against the group's Mong Hsu headquarters on 6 October, days before the signing ceremony. The SSA–N was forced to withdraw from the Tar San Pu river port in Mong Hsu, while the Tatmadaw captured three of the group's outposts on 18 October, before accusing it of having built a new base in violation of a bilateral ceasefire agreement. The fighting led to the displacement of around 6,000 people in Kyethi and Mong Hsu, prompting the United Nations to request that the government allow humanitarian organisations greater access to the conflict zone. In November, some of the camps sheltering displaced persons began to run short on food, and were hit by an outbreak of cholera. The United States called for an investigation into alleged human-rights violations committed by the Tatmadaw during the fighting. On 8 December, following negotiations in Naypyidaw, the SSA–N agreed to withdraw north of a contested highway in Shan State – a demand that it had previously rejected – while the military stated that it would withdraw from some of its positions, and would refrain from establishing new bases.

The previous month, the Tatmadaw had deployed artillery, fighter jets and helicopters to attack KIA camps in Mohnyin township, in Kachin State. The group announced that the offensive was unprovoked, and had forced it to abandon three bases. In November and December, the TNLA engaged in a series of battles in northern Shan State against a force that, unusually, included fighters from both the Tatmadaw and the SSA–S. The latter group denied that it was assisting the Tatmadaw, and accused the TNLA of encroaching on its territory. The TNLA, for its part, stated that the Shan State Army was taking advantage of its signatory status.

Hostility towards Muslims

Myanmar's ethnic minorities were threatened not only by the continued fighting, but also by political developments. This has so far had a limited effect on conflict dynamics, but further violence against the country's Rohingya Muslims is not unlikely. Hostility towards Rohingya increased in the run-up to the elections, and most of them were prevented from voting by changes to electoral regulations. The Union Election Commission also removed around 124 Muslim candidates, some of them Rohingya, from the final electoral list, after their citizenship was called into question (11 of the candidates were reinstated following an appeal).

Against this backdrop, Naypyidaw faced diplomatic pressure over the escalating migration crisis in the Bay of Bengal and the Andaman Sea, across which thousands of Myanmar's Rohingya attempted to flee, often in dilapidated boats operated by human traffickers. As was argued by US President Barack Obama,

Malaysian leaders and others, the government's treatment of Rohingya was a major cause of the crisis. Nonetheless, it continued to insist that they, whom it referred to as 'Bengalis', were not from Myanmar. Partly as a consequence of this, US sanctions against the country were extended on 15 May. Two weeks later, the UN Security Council held its first briefing on human rights in Myanmar. Shortly thereafter, the UN Human Rights Council adopted a resolution condemning the 'systematic gross violations of human rights and abuses committed in Rakhine State, in particular against Rohingya Muslims'.

In 2015, Myanmar's parliament passed the Race and Religion Protection Laws, as advocated by the Committee to Protect Race and Religion, also known as Ma Ba Tha, a nationalist group led by Buddhist monks. The legislation included provisions on monogamy, interfaith marriage, religious conversion and population control, and received heavy international criticism for appearing to target the country's Muslim minority.

According to a report published on 29 October by Yale Law School and non-governmental organisation Fortify Rights, there was 'strong evidence that genocide is being committed against Rohingya people in Myanmar'. Naypyidaw rejected the allegation. A few days later, UN Secretary-General Ban Ki-moon expressed his 'deep concern' about hate speech and the incitement of communal violence by 'extreme elements among the majority community in Myanmar' in the lead-up to the general election.

While the Tatmadaw released around 150 children and young people it had recruited as soldiers, Amnesty International called on Myanmar's authorities to free the 91 political prisoners it believed were still imprisoned in the country. There was hope that the peace process would improve the lives of civilians living in conflict-affected states – not only by reducing the human-rights abuses to which they were subjected, but also by facilitating economic development. However, the process still faced significant hurdles, and would seemingly take years to complete. Furthermore, the transfer of power had the potential to delay substantive peace negotiations until mid-2016. In the interim, the fighting was likely to continue.

Philippines (ASG)

Key statistics	2014	2015
Conflict intensity:	Low	Low
Fatalities:	150	200
New IDPs:		
New refugees:		

The Abu Sayyaf Group (ASG) remained a threat to security in the southern Philippine provinces of Basilan, Sulu and Tawi-Tawi throughout 2015, but its activities had little effect on the stability of the wider state. Manila maintained its well-established position that, despite having been founded by jihadists, the ASG was a criminal gang with no political credibility. Accordingly, the government's strategy for combating the group focused primarily on security operations in Sulu and Basilan which seemed unlikely to be extended beyond those areas.

Security-led strategy

The Armed Forces of the Philippines (AFP) appeared to have some success in their persistent operations to degrade the ASG. General Gregorio Pio Catapang, head of the AFP, said on 15 May that the group was 'in check' and losing internal cohesion due to these operations, which often involved pursuing fighters who had taken hostages and occasionally led to large-scale clashes with significant portions of the group's purported strength. The fact that the ASG leadership seemed willing to mobilise more than 100 fighters at a time – previously a rare occurrence – suggested that the group perceived the military offensives as a serious threat to both its cohesion and its lucrative kidnapping business. The military's capture of some ASG jungle camps also helped deny the militants refuge in Basilan and Sulu, likely exacerbating these concerns.

On 25 February, around 300 ASG fighters – estimated to be more than half of the group's forces – fought the AFP in Patikul, in Sulu. The military reported in March that 37 militants had been killed and 80 others injured in the ensuing violence in the province. A similar clash took place in Patikul on 9 April, when the army made use of helicopter and artillery support to battle around 250 fighters from the ASG. While the AFP also captured one of the group's facilities for producing improvised explosive devices (IEDs) in Mohammad Ajul, in Basilan, on 14 May, around 700 families were displaced by the resulting clashes. A few days later, the security forces seized more IED components, as well as 13 of the fast boats the militants occasionally used to abduct civilians living in Malaysian territory.

The AFP said in July that it would intensify its 'surgical' operations against the ASG to weaken its foothold in Sulu, where the group held most of its hostages and concentrated the majority of its fighters. The military made good on its word on 19 August, engaging an estimated 200 militants near Indanan, in Sulu, and employing artillery fire to suppress the ASG's reinforcements. One soldier and 25 militants were reportedly killed in the battle, while two coastguard officers who had been kidnapped in May escaped during the fighting.

The ASG attempted to discourage further offensives by conducting attacks on civilians, sometimes using IEDs. On 18 September, in what was perhaps the most serious incident of this kind, the group detonated a bomb on a bus in Zamboanga City, killing one person and injuring 31 others. President Benigno Aquino declared in late November that operations against the ASG would again be stepped up, after it beheaded Malaysian hostage Bernard Then Ted Fen. As a consequence, the military redeployed around 500 troops from Bicol Region to Mindanao, and conducted several major offensives against the ASG the following month. The operations led to the seizure of an ASG camp in Darayan, in northwestern Sulu, on 2 December, and to the deaths of three soldiers and 26 militants shortly thereafter, during a three-day mission to capture a camp near Barka – a Basilan town in which the ASG had a strong presence and local support. The group attempted to hit back on 15 December, deploying an estimated 150 fighters to attack the headquarters of the 32nd Infantry Battalion in Patikul. However, they were repelled by the AFP, and suffered an unknown number of casualties. On 30 December, one soldier and ten suspected members of the ASG were killed in another major firefight in Patikul. The military claimed to have killed a total of 133 Abu Sayyaf militants, and to have lost 18 members of the security forces, in Sulu in 2015. Only 12 of the 133 bodies were recovered by the military, meaning the fatality figures could not be verified.

Aquino had promised 'significant accomplishments' against the ASG in late 2014, stating that developmental efforts were to be part of the strategy for combating the group. Although it was reported in January 2015 that Barka would receive additional funding for development, the government continued to place a strong emphasis on security measures. Given that the deployment of significant military resources was not enough to destroy the ASG, it seemed that greater political efforts would be required to address some of the issues linked to the group's continued activities. These included its support among some local clan networks, the poverty that it could exploit to recruit under-employed young men and the proliferation of small arms in the southern Philippines.

Having sworn allegiance to the Islamic State (also known as ISIS or ISIL), the ASG frequently employed the Middle Eastern jihadist group's signature black flags for propaganda purposes. Yet the government insisted that there was no direct contact, nor any financial links, between ISIS and non-state armed groups in the Philippines – and this claim appeared to be true. The most significant political concern relating to the ASG was rather its connection with similar organisations elsewhere in the Asia-Pacific: five Malaysian militants were seen accompanying ASG fighters in Sulu and Basilan in June; in August, the AFP identified four members of Jemaah Islamiah among ASG forces in Patikul; and, in December, a Malaysian bomb maker was killed during a battles between the military and the ASG.

Malaysia's counter-terrorism police said in November that three Malaysian militants affiliated with ISIS were in the Philippines, looking to establish an official Southeast Asian faction of the Middle Eastern group. If formed, this new branch would, according to the Malaysian counter-terrorism police, incorporate the ASG and other similar organisations. Nonetheless, instead of signalling a qualitative change in its activities, the ASG's pledge of allegiance to ISIS was likely made in a bid to aid promotional and recruitment efforts.

Kidnapping business

The ASG and other kidnapping-for-profit criminal groups (which often sold their victims to the ASG) had a clear impact on locals and tourists in the southern Philippines. Foreign, especially Western, citizens were attractive targets for the ASG, due to their value as a source of propaganda and funding. Partly due to its kidnapping revenues, the group appeared to be relatively well funded. This was reflected in a statement by the mayor of Mohammad Ajul, who claimed that the ASG continued to recruit fighters in the southern Philippines – including Zamboanga City and other urban areas – by offering them as much as US$1,100 per month.

The decapitated body of Rodolfo Boligao, a municipal official from Zamboanga del Norte who had been kidnapped by the ASG in May, was found on 11 August in Maimbung, in Sulu. His family insisted that they had sent the group the ransom to free him through an emissary. Bernard Then, the Malaysian hostage whose execution elicited a response from Aquino, was also killed due to the alleged non-payment of a ransom – although the militants reportedly raised the amount they demanded for him more than tenfold. Another Malaysian abducted at the same time was released, apparently after the ransom was paid. Following the incident, and the death (by illness) of a South Korean abductee in October, Seoul banned its citizens from travelling to Basilan, Sulu and Tawi-Tawi due to the high security risk.

The ASG demanded US$21 million for each of four hostages abducted in Davao del Norte on 21 September: two Canadians, one Norwegian and one Filipina. In a video featuring the victims published on 12 October, the group also demanded a halt to the military campaign against it. The four captives remained in the hands of the ASG at the end of the year, despite the security forces' attempts to rescue them.

Civilians were also affected by the group's other efforts to gain control in local communities, as well as the government's counter-terrorism operations. The ASG was suspected to have killed a village leader and his aide in Sumisip, in Basilan, on 11 April, and to have kidnapped and beheaded a civilian in Mohammad Ajul the following month. In the latter incident, the group mistook the victim for a military informant. Around 27,000 people temporarily lost access to their water supply on 6 June, when ASG fighters bombed a pipe outside the town of Maluso, in Basilan. On 23 June, the group detonated an IED in front of a remittance centre in Lamitan, in Basilan, although no one was injured. Meanwhile, operations conducted by the military in December displaced around 1,000 families in Basilan. By the end of the year, the conflict appeared no closer to a conclusion, despite the AFP's efforts, and the ASG is likely to remain a threat until governance and rule of law improves in the affected areas.

Philippines (MILF)

Key statistics	2014	2015
Conflict intensity:	Low	Low
Fatalities:	150	300
New IDPs:	95,000	
New refugees:		

Efforts to end the conflict between the Moro Islamic Liberation Front (MILF) and the Philippine government were severely hampered by two factors in 2015. The first was the killing of dozens of police officers in Mamasapano, in Maguindanao, during a disastrous security operation, while the second was a presidential election campaign that coincided with the final stages of the peace process between the sides. As a result, the Philippine Congress missed the deadline for passing the Bangsamoro Basic Law (BBL), which President Benigno Aquino had put forward to establish a Bangsamoro autonomous region before 2016. A key provision of the peace agreement, the BBL was therefore unlikely to become law during Aquino's tenure, increasing the risk of a return to insurgency by disgruntled insurgents.

Mamasapano and its aftermath

The Special Action Force (SAF), a unit of the Philippine National Police, conducted on 25 January an operation to capture or kill Zulkifli Abdhir, a Malaysian bomb maker who worked with MILF splinter group the Bangsamoro Islamic Freedom Fighters (BIFF) and Islamist militia the Abu Sayyaf Group. Due to concern about leaks, however, the SAF had not coordinated the operation with the nearby 105th MILF Command, which believed it was being targeted. As a consequence, 44 members of the SAF were caught in an open field and killed by MILF and BIFF fighters. Abdhir, 17 MILF fighters and four civilians also died in the battle. The police were unable to kill their secondary target, bomb maker Abdul Basit Usman. Although Philippine soldiers nearby were aware of the operation, they did not provide artillery support to the SAF due to a reluctance to break the ceasefire with the MILF. The government's Coordinating Committee on the Cessation of Hostilities subsequently stated that the SAF routinely ignored the rules of engagement in MILF-controlled areas, and in doing so endangered the peace process.

The killing of so many police officers had immediate political repercussions. Two senators withdrew their support for the BBL the day after the incident, with the result that the law lost its automatic majority in the Senate. Many Filipinos were incensed by the fact that Abdhir and Usman had been given refuge in what was thought to be MILF-controlled territory, and public discontent grew after the emergence of a video that showed unidentified militants executing injured officers and taking their equipment. Opposition politicians took the opportunity to attack Aquino, while the MILF ascribed the incident to 'orchestrated spoiling'. Meanwhile, General Gregorio Pio Catapang, chief of the army, attempted to reduce tensions, stating that a resumed war with the MILF would be 'absurd' and 'illogical'.

Separate investigations into the Mamasapano operation were launched by the government, the MILF and the International Monitoring Team, which oversaw the ceasefire between the sides. The MILF's report blamed a lack of coordination by the government, stating that the security forces violated the ceasefire by shooting first, and that an SAF officer had murdered four MILF fighters as they slept. The group denied that its members had desecrated the bodies of the police officers, and claimed to have been unaware that Abdhir and Usman were in the area – although, on the latter point, it recommended that two of its own commanders be disciplined for a 'failure of intelligence'. Finding no evidence that the MILF knew the bomb makers were in its territory, the International Monitoring Team blamed both sides, but primarily the government, for violating the ceasefire. The government report said the operation was poorly planned, and criticised commanding officers, as well as President Aquino, for not properly following the chain of command.

The proximity of the 105th MILF Command to the BIFF group also came under scrutiny, with a spokesperson for the latter saying that they were 'like family'. This suggested the MILF's central leadership had limited control of some fighters on the ground, perhaps partly due to clan loyalties in Mindanao. Indeed, despite the peace agreement, hundreds of families had been displaced on 17 January, when a *rido* (clan feud) apparently led members of the 105th to attack a village in Shariff Aguak, in Maguindanao.

Following the Mamasapano incident, the Armed Forces of the Philippines (AFP) and the MILF continued, and possibly strengthened, their coordination on the battlefield. The military assisted on 9 February a MILF unit under attack from Abu Sayyaf fighters in Basilan Province, and the MILF subsequently ordered its units to stand aside during an AFP offensive against the BIFF. Furthermore, the MILF attacked BIFF units in Maguindanao, and stated that it would help hunt down Usman through the Ad Hoc Joint Action Group, a mechanism in which the government and the MILF cooperated to 'interdict and isolate' terrorists and other criminals. On 18 May, the MILF joined UNICEF in launching a programme to prevent the recruitment of child soldiers in Moro communities, becoming the world's only non-state armed group to have participated in such an effort.

The killings in Mamasapano also led many politicians to perceive support for the BBL as potentially damaging to their popularity. In light of this, some members of congress joined campaigners in describing the bill as unconstitutional, and calling for it to be amended. Some of the most controversial provisions of the law related to its creation of a police force for Bangsamoro, and to the proportion of revenue from natural resources allocated to the autonomous region. Miriam Coronel-Ferrer, the government's chief peace negotiator, blamed the delay in passing the bill on Aquino's political opponents, whom she accused of spreading misinformation on the Mamasapano incident.

The MILF declared that it would accept some amendments to the BBL, as long as they were in line with the 2014 Comprehensive Agreement on Bangsamoro. For example, the group opposed the House of Representatives' removal of a provision allowing areas contiguous with Bangsamoro to become part of the region if the majority of their inhabitants voted in favour of the move. The provision also faced resistance from residents of these areas: there were reports that influential locals in Wao, in Lanao del Sur, instigated violence to split communities along ethno-religious lines, with the aim of preventing the territory from becoming part of Bangsamoro.

A senior MILF representative criticised the energy and mining lobbies in Congress for attempts by members of the Congress to designate nearly all mineral and energy resources in Mindanao as 'strategic'. This designation would cut off the regional government's access to revenues from the industries – in contravention of one of the MILF's key demands. The group warned Congress not to dilute the peace agreement, but stated repeatedly that it would not withdraw from the peace process, regardless of whether the BBL was voted into law. Nonetheless, MILF negotiator Mohagher Iqbal said that Congress's failure to pass the BBL could portray his organisation as weak, strengthening more radical Muslim groups that opposed the peace process. The Organisation of Islamic Cooperation also warned of a possible spread of radicalisation if the deal were to fail; on 7 June,

137 Bangsamoro community leaders asked the MILF to revert to its original goal of independence if Congress were to pass a watered-down version of the BBL.

The group attempted to sustain the peace process, reportedly killing Usman on 3 May. In the wake of his death, Catapang declared that peace in Mindanao had become more attainable, while the president's office said it signalled the MILF's reliability as a peace partner. A more cynical theory was that Usman was sacrificed to prevent MILF commander Wahid Tundok from facing criminal charges in relation to Mamasapano.

Despite the delay in passing the BBL, the MILF began on 16 June to decommission weapons and demobilise fighters. At a ceremony in Maguindanao attended by President Aquino, the group handed over 20 high-powered and 55 crew-served weapons, while 145 of its members began the reintegration process. Presidential peace adviser Teresita Deles announced that the government would spend US$53 million on supporting former fighters in 2015.

One of the leaders of the campaign to amend the BBL was Senator Ferdinand 'Bongbong' Marcos Jr, son of former president Ferdinand Marcos. The senator submitted a substitute bill, the Basic Law of the Bangsamoro Autonomous Region (BLBAR), to the Senate on 10 August. Around 80% of the BBL's provisions were amended by the new bill – most controversially, those relating to a US$360m development fund for the region, which was scrapped, and to sharing revenue from natural resources. The MILF viewed the BLBAR as providing less autonomy than the agreement it was intended to replace, and Iqbal declared that he would ask Congress to reinstate 28 passages that had been removed from the BBL, including its preamble and a provision on revenue from natural resources. A separate revised version in the House of Representatives removed 48 'unconstitutional' provisions, according to Representative Rufus Rodriguez. House Speaker Feliciano Belmonte contended that the provisions removed by the House would not be restored, and Marcos argued that the original BBL had 'no chance of passing in the Senate'. The two houses of Congress repeatedly filibustered, and failed to achieve a quorum for voting on, the heavily amended versions of the bill.

MILF chairman Murad Ebrahim announced on 31 August that the BLBAR was a 'complete violation' of the peace agreement. He added that the MILF had its 'own plan' if the final BBL did not comply with the pre-existing agreement, and that the group was considering both political action and an armed struggle. In an 8 September editorial published on its website, the MILF stated that it would reject a diluted BBL outright, and that 'various aspects of the normalization process including decommissioning of its weapons (and combatants) will come to a halt'. Two weeks later, Philippine prosecutors investigating the Mamasapano incident filed assault and murder charges against 90 people, 26 of whom were members of the MILF.

Despite the setbacks, the MILF worked to increase support for the peace process among other militant groups. On 13 October, the MILF and the faction of the Moro National Liberation Front (MNLF) led by Muslimin Sema made a 'unified declaration' expressing their support for an undiluted version of the BBL. Even more significantly, the hardline faction of the MNLF led by Nur Misuari, which had opposed the BBL as a violation its own 1996 peace agreement, changed its stance to non-opposition following meetings with the MILF in November.

Meanwhile, the BIFF continued to oppose the peace process and carried out several terrorist attacks, perhaps sensing a chance to further damage the relationship between the MILF and the government. More than 100 fighters, or an estimated half of BIFF members, were killed during the AFP's offensive against the group in central Mindanao in February and March. By 20 March, the campaign had displaced around 100,000 people in Maguindanao, half of whom returned home within a month. Another 20,000 people had been displaced by fighting between the MILF and the BIFF in North Cotabato and Maguindanao in February. The BIFF was significantly weakened on 14 April, when its founder and leader, Ameril Umbra Kato, died of natural causes. The group appointed Esmael Abubakr as its new leader, and continued to carry out hit-and-run attacks, but on a much reduced scale.

Seeking to build on its gains, the AFP declared on 8 April that it was deploying additional soldiers to Maguindanao villages that had been cleared of BIFF fighters. There, it aimed to 'establish a government presence' and to prevent the return of the rebels. Additionally, an infantry battalion was relocated from Eastern Visayas to Maguindanao to support operations against the BIFF, the United Justice Movement and other 'lawless elements'. Following what it described as a 'post-conflict needs assessment', Manila also approved construction of a Marine Corps base in Rio Hondo, in Zamboanga City, to help improve security in surrounding areas. Furthermore, the government ordered the creation of a National Task Force for the Disbandment of the Private Armed Groups, based primarily in Bangsamoro. Such groups, and the prevalence of clan warfare, continued to be persistent security threats in the region.

The BIFF stepped up its terrorist attacks in December, killing eight civilians and ending a period of relative stability that had lasted for eight months. BIFF spokesman Abu Misry Mama stated that the assaults demonstrated that the group had retained its capabilities and continued to oppose the presence of government forces in Bangsamoro. He added that the operations were inspired by Middle Eastern jihadist group the Islamic State (also known as ISIS or ISIL). The attacks indicated that the BIFF was eager to demonstrate its continued relevance.

In November, the first batch of 75 MILF fighters began training to become part of a Joint Peace and Security Team, which would also consist of soldiers and police officers. Conducted in preparation for the potential transition to an autonomous Bangsamoro region, the training was held in Mindanao by the AFP and Philippine National Police. Australia pledged AU$4m to support the justice and joint-security mechanisms of the MILF–Manila peace process over the three years. Yet the success of such efforts depended on Congress passing the BBL. The task of completing the peace process was likely to pass to the next president, who would not necessarily share Aquino's determination to see it through. There was also a risk that disgruntled MILF fighters would again take up arms against the government, and perhaps join the BIFF, if no satisfactory compromise was reached.

Philippines (NPA)

Key statistics	2014	2015
Conflict intensity:	Low	Low
Fatalities:	150	250
New IDPs:		
New refugees:		

There was no change in the drivers and dynamics of the conflict between the Philippine government and Maoist insurgent group the New People's Army (NPA) in 2015. Although the sides attempted to pursue a peace process, back-channel 'talks about talks' collapsed amid mutual recriminations, and the annual fatality rate rose by two-thirds. The NPA persisted in its guerrilla campaign against the security forces – relying primarily on ambushes and the use of improvised explosive devices (IEDs) – while competing with the government for the support of indigenous people, the main victims of the violence.

Manila and the Communist Party of the Philippines (CPP), the NPA's parent political organisation, both expressed a willingness to discuss peace, but doubted the sincerity of the other side. Luis Jalandoni, chairman of a peace panel set up by the National Democratic Front of the Philippines (NDFP) – an umbrella group of leftist organisations controlled by the CPP – declared that the militant group was willing to talk with the government at any time. However, he also accused President Benigno Aquino of obstructing the peace talks by focusing on a separate peace process with the Moro Islamic Liberation Front (MILF).

On the 42nd anniversary of the NDFP, founder Jose Maria Sison claimed that the NPA had won 'brilliant victories on the battlefield' and was on the verge of achieving its medium-term goal, a strategic stalemate with the armed forces – although this view was hard to reconcile with the group's losses on the ground. He added that the NDFP 'continues to explore' peace negotiations, and that the prosecution of a 'protracted people's war' was the communist movement's most important task. Aquino announced that the NPA would have to demonstrate its sincerity before talks could resume. Yet senior communist leaders accused the president of sabotaging the peace process, and demanded that the government comply with existing accords while putting forward a 'comprehensive agreement on social and economic reforms'. Given that the authorities occasionally arrested the NPA's 'peace consultants', the existing accords in question were likely those of the 1995 Joint Agreement on Safety and Immunity Guarantees, which stipulated that peace negotiators not be apprehended.

Following a meeting with House Speaker Feliciano Belmonte and Jalandoni in the Netherlands, Sison said in June that informal peace talks could begin sometime between July and September. The effort fell apart almost immediately, however, as Jalandoni insisted that the process could not begin until the government released the NPA's peace consultants and 'political prisoners'. When presidential peace adviser Teresita Deles pointed out the difficulty of meeting these conditions, Jalandoni accused her of engaging in an 'extremely malicious kind of preconditioning'. As a consequence, the armed conflict continued.

Continued insurgency

In a series of announcements that continued throughout the year, the Armed Forces of the Philippines (AFP) claimed to have strengthened security in various provinces and towns, suggesting that the insurgency had diminished. But the NPA remained a threat, and the AFP's declarations of victory were not always accompanied by a decline in militant activity, and conflict-related fatalities increased by about 100.

Having declared six provinces 'insurgency-free' in January, the AFP announced on 6 April that the second and third districts in Negros Occidental Province were 'peaceful and ready for further development'. This was followed on 23 April by the announcement that Albay Province had become a 'manageable conflict area' containing an estimated 40 NPA fighters, and that another nine towns in Leyte Province and three in Samar Province were insurgency-free and ready for development.

The insurgents were weakened considerably in Davao Region on 28 June, when Leoncio Pitao, the most-wanted NPA commander in Mindanao, was killed in a clash with security forces in Davao City's Paquibato district. Davao City Mayor Rodrigo Duterte called for a temporary ceasefire during Pitao's wake, to allow NPA fighters in mourning to enter the city. Duterte called Pitao a 'friend', while urging the NPA to refrain from revenge attacks.

Meanwhile, government forces continued to conduct offensives near Kitaotao, in Bukidnon Province, leading to several large-scale battles. The AFP recovered large amounts of weapons, IEDs and bomb-making material at an NPA base in White Kulaman village on 28 August, before capturing another of the group's camps two days later.

According to the military, 340 members of the NPA were 'neutralised' (killed or captured) in the first half of the year, leaving the group with around 4,100 fighters – significantly more than the previous estimate of 3,200, made in January. Yet a presidential spokesperson claimed in December that the number of NPA troops had fallen from more than 2,000 in April to 1,691 in June. He also stated that, in the same period, there had been a 25% decrease in wards affected by NPA activity (from 547 to 414), as well as a reduction in both the fronts on which the group was fighting (from 29 to 24) and the number of weapons it possessed. The NPA claimed that it was active on 46 fronts, however, and that it had acquired a 'mass base' of 200,000 supporters. The group also declared that it had opened five fronts and acquired nearly 1,000 high-powered firearms during Aquino's presidency. Although none of the claims could be verified, the NPA likely retained around 4,000 fighters – roughly the same number that it had had since 2010.

The group continued to carry out assaults on businesses, which it accused of human-rights violations but appeared to have targeted due to their failure to pay 'revolutionary taxes', or protection money. The NPA reportedly collected around US$50,000 per month from firms in Compostela Valley Province. Much of this funding was taken from mining and agricultural businesses. The NPA was also criticised for extorting politicians, who were sometimes forced to pay the group for a 'permit to campaign' or a 'permit to win'.

Several NPA fighters surrendered in the year, aiming to join disarmament, demobilisation and reintegration programmes. The group targeted those who had left, even if they had done so decades earlier, killing on 7 June a man who had surrendered in 1986. In October, the NPA abducted and killed Dario Otaza – mayor of Loreta, in Agusan del Sur Province – as well as his son. Otaza was an indigenous Lumad who had also left the NPA in 1986 and who was thought to have helped more than 100 of the group's fighters join the rehabilitation programme. While the NPA justified the killings on the grounds that Otaza was 'terrorising' indigenous people, Human Rights Watch described its 'revolutionary justice' as 'just plain murder', adding that the group's actions were 'flagrant violations of international law'.

Damage to indigenous communities

With the Independent Commission for Human Rights accusing both the AFP and the NPA of severely mistreating civilians, the sides traded accusations over which had caused greater harm to Lumads in Mindanao. Lumad civil-society organisations responded by denouncing the claims as propaganda, observing that indigenous people were often incorrectly labelled as NPA members or military informants. Colonel Eduardo Gubat, spokesperson for the AFP's Eastern Mindanao Command, caused an uproar by arguing that a group of Lumads sheltering in a Davao City church had been manipulated by human traffickers rather than displaced by the conflict. Gubat had based his assessment on a report by Chaloka Beyani, UN special envoy for internally displaced persons. Yet, in the report, the Lumads in question stated that they were most concerned about the military and paramilitary forces in their communities. Due to his error, Gubat was forced on 13 August to resign as spokesperson, and to apologise for misrepresenting Beyani.

Meanwhile, other Mindanaoan civil-society organisations called for an end to the government's counter-insurgency campaign, arguing that the AFP had, inter alia, forcibly recruited indigenous people, carried out extrajudicial killings and helped private companies pillage the country's resources. However, such groups were accused by the military of having links to the NPA, and the military denied all the accusations.

Magahat–Bagani, a militia allegedly affiliated with the government, was thought to be responsible for several attacks on civilians, including an assault on 1 September in Lianga, in Surigao del Sur Province, in which three people were killed. Johnny Pimentel, governor of the province, subsequently demanded that Magahat–Bagani be disarmed, disbanded or destroyed to protect indigenous communities, contending that the militia was 'under the control and supervision of the Army'. Two UN special rapporteurs called for an independent investigation into the killings, while the Philippine police and military announced the formation of a joint task force to tackle the group.

Human-rights group Karapatan, also linked to the NPA by the military, declared on 6 September that more than 2,700 indigenous people had been forced from their homes by security operations in Surigao del Sur. Shortly afterwards, around 60 tribal leaders in Agusan del Sur called on the NPA to leave their ancestral domains, and announced that the group was primarily responsible for the killing of indigenous people in their territories. This hostility towards the insurgents stemmed partly from their manufacture of explosives and use of IEDs in these areas.

The armed forces continued to demand that the NPA end its recruitment of children – an issue that came to the fore on 9 May, when an 11-year-old member of the group surrendered to soldiers in Asuncion, in Davao del Norte. According to one NPA leader, around 22% of the group's members were between six and 18 years of age.

The threat to civilians became particularly apparent in early October, when fighting in Mahaba, in Surigao del Sur, displaced around 1,000 indigenous people. While the NPA was far from securing its strategic stalemate, the AFP appeared unable to defeat the group outright. The 30-year-old peace process therefore seemed the best means for ending the conflict. But both sides lacked the will to commit to the effort, ensuring that civilians would continue to be caught up in the violence.

Southern Thailand

Key statistics	2014	2015
Conflict intensity:	Medium	Medium
Fatalities:	350	250
New IDPs:		
New refugees:		

The insurgency in southern Thailand diminished slightly in 2015, as the government made a renewed effort to reach a peace agreement with the myriad militant groups in Narathiwat, Pattani and Yala provinces. The dynamics of the conflict changed little, and its annual fatality rate fell to around 250, although the insurgents continued to carry out indiscriminate attacks on the security forces, anyone perceived as representing the state – including teachers – and other civilians. While most of their assaults involved drive-by shooting, ambush or the detonation of an improvised explosive device (IED), they also conducted coordinated operations that employed large numbers of bombs.

Stuttering peace process

In January, the Pattani United Liberation Organisation (PULO) attempted to form a coalition with its fellow insurgent groups, in line with Bangkok's stated preference for negotiating with a united front. To this end, the organisation held a meeting in Germany that included Barisan Revolusi Nasional (BRN), Barisan Islam Pembebasan Pattani (BIPP) and Gerakan Mujahidin Islam Pattani (GMIP). In March, three PULO factions, the BIPP and the GMIP joined Mara Patani (also known as the Patani Consultative Council), an umbrella organisation created in November 2014 by pro-dialogue members of the BRN, the largest of the groups. Given that the members of Mara Patani had credibility in the fragmented insurgent movement, the development appeared to be a step towards peace.

However, on 20 February, 17 civilians were injured by a car bomb in Muang district, in Narathiwat Province. The fact that the attack occurred during Lunar New Year week led Thai officials to speculate that the insurgents were targeting the tourism economy. Yet it seemed possible that the assault had been spurred by dissatisfaction with the peace process. Likewise, a large-scale bombing spree in Yala in May, and the detonation of a car bomb by suspected insurgents on the tourist island of Ko Samui in April, suggested that some insurgents were uninterested in talks with the government. On 19 May, cloth banners denouncing leaders in Bangkok as 'colonialist oppressors' were found in Narathiwat's Muang, Sungai Padi and Bacho districts. Both the attack in Ko Samui and another in Songkhla Province in July were unusual in that they took place outside the insurgents' normal area of operations. The assaults in Yala, which involved around 40 bombs and injured dozens of people, likely required the kind of coordination only possible with the participation of senior rebel leaders.

Meanwhile, the armed forces made few changes to their security measures in the south. The army stated its intention to expand the 'Thung Yang Daeng' model of popular surveillance across the region, establishing communication centres in nearly 300 sub-districts to enable locals to report militant activity. The model emphasised the enlistment of civilians in the counter-insurgency effort, adding to existing schemes such as the 'Chor Ror Bor', which centred on locals given limited combat training and armed with government-issue weapons. Furthermore, on 21 March, the governor of Narathiwat ordered the installation of 600 security cameras in the province to help prevent insurgent attacks. Northern troops deployed to the south would be withdrawn from April onwards, due to an improvement in the security situation. The Internal Security Operations Command (ISOC) stated that northern army regiments stationed in the south would be replaced with soldiers from a Yala-based regiment, in a process that could take up to three years. In the interim, the shortfall in troop numbers would be compensated for by local volunteers.

The government announced on 25 June that it had held talks with Mara Patani. But the peace process made only limited progress, due to the sides' incompatible negotiating positions and opposition to the undertaking from some insurgents. National Security Council Secretary-General Anusit Kunakorn argued that the talks should be expanded to include rebel groups outside the coalition, whom he blamed for a series of bombings that killed six people in July. That attack prompted the authorities to improve security along the Sungai Kolok river, near Malaysia, after evidence emerged that the explosives used by the perpetrators had been smuggled across the border.

The same month, the government attempted to boost the peace process by releasing Sama-ae Thanam, former head of the PULO, after he had served 18 years of a life sentence. Upon his release, Sama-ae said he would be willing to help coordinate talks, although he later revealed that he had turned down an invitation to do so. Two months later, Bangkok permitted the release of Da-Oh Thanam, another former leader of the PULO. However, neither gesture appeared to have its intended effect.

On 9 October, Bangkok presented Mara Patani with a proposal for a peace plan centring on three points: the establishment of safety zones in areas affected by the conflict, the creation of means for addressing urgent issues and the formation of an alternative justice system. Yet the coalition denied having agreed with the government to establish safety zones in two Narathiwat districts.

Representatives of the BRN subsequently rejected the peace process as a 'negative phenomenon' in which their group was 'categorically not involved', and suggested that it failed to comply with 'international standards and norms'. They insisted that the government needed to involve countries other than Malaysia in peace mediation. Thailand's chief negotiator, General Aksara Kerdphol, commented that the BRN was split over how to approach the talks, and a spokesman for the ISOC urged the group to join the negotiations rather than criticise them. The secretary of the Thai negotiating team, Lieutenant-General Nakrob Boonbuathong, described the BRN's statements as a public-relations strategy, and claimed that the group's delegates in Mara Patani had confirmed their participation in the talks. Nonetheless, negotiations initially scheduled for November were postponed until 2016.

A series of 35 coordinated bombings and arson attacks took place across Yala Province on 29 October. The assaults caused no casualties, as they primarily targeted electricity pylons. Nine days earlier, three people had been injured when six bombs were detonated in Khok Pho district, in Pattani. The ISOC approved on 21 December the withdrawal of around 1,500 security personnel from conflict-affected areas in the south, leaving around 31,000 soldiers, 18,000 police officers and 9,600 defence volunteers in the region. The move followed the Royal Thai Air Force's deployment of 287 personnel to Bor Thong airport, in Pattani's Nong Chik district, to enhance security there and at other locations in the south.

Pervasive human-rights violations

The security forces continued to operate with impunity in the south, due to the government's imposition of martial law (revoked elsewhere in the country on 1 April). As part of its emergency powers under Section 44 of the interim constitution, the army was able to ban media reports, arrest suspects for seven days without a warrant and carry out raids as deemed necessary. 'Political gatherings' of more than five people were also banned.

On 2 April, the police arrested 22 students who had gathered to demand an investigation into an incident the previous week, in which the Pattani security forces had killed four civilians after mistaking them for insurgents. General Prawit Wongsuwan, Thailand's deputy prime minister, announced that the officers responsible would 'undergo the judicial process', and that the commander and deputy commander of the regiment involved in the incident had been transferred to Songkhla. The PULO strongly criticised the government over the killings.

On 21 August, a Songkhla court ordered the prime minister's office to pay compensation to the family of a man who was beaten to death in custody in 2007. No security forces were found guilty of wrongdoing, however. One month later, the National Anti-Corruption Commission declared that it would ask prosecutors to take action against an army officer who had allegedly tortured an imam to death in Narathiwat in 2008. On 4 December, a man held at Inkhayutthaborihan camp, in Nong Chik, died in military custody after reportedly admitting to being an insurgent during interrogation. He was the third suspect to die at the facility since October. After carrying out an autopsy, however, the police stated that his body showed no signs of injury.

Forced DNA testing also exacerbated the sense of alienation evident among many residents of the south. The police announced on 7 May that they had taken DNA samples from 40,000 people in southern provinces, with a view to investigating insurgent attacks. The security forces even gathered DNA samples from a baby without his parents' permission, while conducting a raid in Nong Chik on 21 November – an incident that sparked demonstrations. The baby's father, who was absent during the raid, was reportedly a suspected insurgent.

In response to the rebels' indiscriminate attacks on civilians, the government banned the steel cooking-gas cylinders they often used to create IEDs in Yala, Pattani, Narathiwat and four districts of Songkhla (the steel cylinders were replaced by those made of composite material). The move had the potential to reduce the number of casualties from IED attacks, if carried out effectively. Thailand's and Malaysia's police forces also began exchanging intelligence on thefts of motorcycles, often used to deliver IEDs, in areas near the border between the countries.

In summer, a southern teachers' organisation demanded that those it represented receive enhanced protection by the security forces. This followed the insurgents' murder of a teacher on 26 May – the 182nd such killing during the conflict, according to the organisation. However, another teacher was killed in a drive-by shooting in Muang on 13 July, while four rangers, five children and another civilian were injured when a bomb exploded near the entrance to a school in Yarang district on 11 September. These attacks occurred despite the fact that the security forces already provided an armed escort to teachers travelling to and from work. The insurgency also affected children in other ways: Child Soldiers International and the Cross Cultural Foundation reported that rebel groups in the south had recruited civilians as young as 14.

Chapter Seven

Europe and Eurasia

Armenia–Azerbaijan (Nagorno-Karabakh)

Key statistics	2014	2015
Conflict intensity:	Low	Low
Fatalities:	55	60
New IDPs:		
New refugees:	1,250	

Throughout 2015, the conflict in Nagorno-Karabakh was characterised by active but mostly ineffective diplomacy, conducted against a backdrop of escalating violence and belligerent rhetoric from both sides. Although the security situation in the conflict zone showed clear signs of deterioration, there was little change in the political positions of Armenia and Azerbaijan, or in the drivers and dynamics of their interaction. The countries began once again to accuse each other of spreading disinformation, making it difficult for outside observers to analyse the scope of violent incidents and their impact on civilians.

Moreover, two worrying trends emerged in the year: the sides increasingly relied on heavy weaponry, and incidents were reported in previously uncontested areas near the Armenia–Azerbaijan border.

The parties released in late March contradictory reports on several large-scale battles. The Azerbaijani Ministry of Defense claimed that up to 20 fighters from Nagorno-Karabakh had been killed after attacking Azerbaijani forces near the

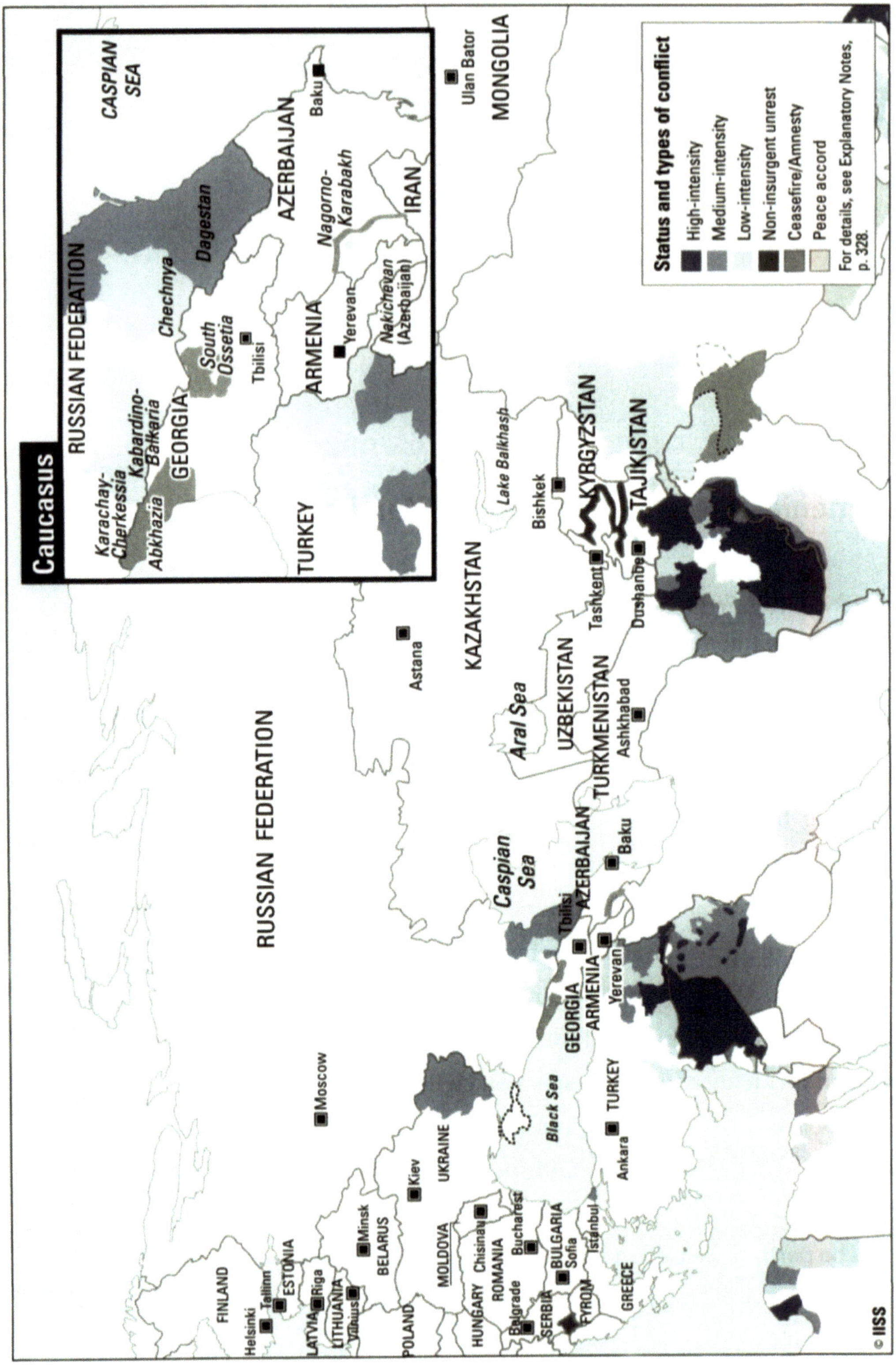
Caucasus
RUSSIAN FEDERATION
CASPIAN SEA
Karachay-Cherkessia
Kabardino-Balkaria
Abkhazia
GEORGIA
Chechnya
Dagestan
South Ossetia
Tbilisi
AZERBAIJAN
Baku
ARMENIA
Yerevan
Nagorno-Karabakh
Nakichevan (Azerbaijan)
TURKEY
IRAN
RUSSIAN FEDERATION
Moscow
Astana
KAZAKHSTAN
Lake Balkhash
Ulan Bator
MONGOLIA
Bishkek
KYRGYZSTAN
TAJIKISTAN
Tashkent
Dushanbe
UZBEKISTAN
TURKMENISTAN
Ashkhabad
Aral Sea
Caspian Sea
Baku
AZERBAIJAN
Tbilisi
GEORGIA
ARMENIA
Yerevan
TURKEY
Ankara
Black Sea
UKRAINE
Kiev
BELARUS
Minsk
MOLDOVA
Chisinau
ROMANIA
Bucharest
BULGARIA
Sofia
Istanbul
GREECE
FYROM
SERBIA
Belgrade
HUNGARY
POLAND
LITHUANIA
Vilnius
LATVIA
Riga
ESTONIA
Tallinn
FINLAND
Helsinki
Status and types of conflict
High-intensity
Medium-intensity
Low-intensity
Non-insurgent unrest
Ceasefire/Amnesty
Peace accord
For details, see Explanatory Notes, p. 328.
© IISS

Line of Contact. Yet sources in the Nagorno-Karabakh Defense Army claimed that Azerbaijani troops had attacked its positions. According to the Nagorno-Karabakh authorities, three Nagorno-Karabakh soldiers had been killed and four others wounded, while between seven and 14 Azerbaijani soldiers had been killed. Shortly afterwards, the Azerbaijani Ministry of Defense claimed that its Armenian counterpart had covered up the deaths of five Nagorno-Karabakh fighters in a separate incident. In response, the spokesperson for the Armenian Ministry of Defence said that both it and the Nagorno-Karabakh Defense Army were fully transparent, and called on Armenian media outlets to refrain from repeating 'anti-Armenian propaganda'.

Azerbaijani servicemen and troops from the Nagorno-Karabakh Defense Army clashed in the Tetersky region on 21 April. The Azerbaijani Ministry of Defense reported that five Nagorno-Karabakh troops were killed and another ten injured after they opened fire on Azerbaijani positions. But the Nagorno-Karabakh authorities declared that they had not lost any troops, while one Azerbaijani soldier had been killed, adding that the 'information circulated by the Azeri side is just disinformation aimed at hiding its own casualties.'

A spokesperson for the Armenian Ministry of Defence reported in June that one of its soldiers had been killed outside the town of Gorhayk, in an area of Armenia bordering the Azerbaijani exclave of Nakhchivan. He commented that Azerbaijani forces had opened fire without warning, prompting the Azerbaijani Ministry of Defense to deny the accusation (without providing further details on the battle).

Fighting along the Line of Contact intensified between July and September, as both sides began to report battles in previously uncontested areas. Throughout the summer, the Nagorno-Karabakh authorities publicly accused Azerbaijan of firing rounds from mortars and heavy weapons across the Line of Contact. The escalation there peaked between late August and late September. Nagorno-Karabakh officials reported on 24 August that Azerbaijani forces had shelled the positions of Nagorno-Karabakh soldiers. After launching a counter-attack, Nagorno-Karabakh forces claimed to have killed four Azerbaijani soldiers and wounded more than 15 others. On 25 September, Azerbaijani troops reportedly used Turkish-made TR-107 rocket launchers to target Nagorno-Karabakh forces. A few days later, the sides clashed once more. They again released conflicting casualty reports: the Azerbaijani Defense Ministry said that three of its troops and seven from Nagorno-Karabakh had been killed, while the Nagorno-Karabakh authorities claimed that more than ten Azerbaijani soldiers had been killed or wounded.

Following the surge in violence that began in late August, the US Department of State called for 'an immediate end to such violence, for the de-escalation of tensions, and, in the strongest possible terms, for the avoidance of civilian casualties'. According to the US, the sides' use of heavy weapons in early September had caused civilian casualties and damage to infrastructure in villages near the Armenia–Azerbaijan border.

Meanwhile, Azerbaijani and Armenian troops clashed along the northern part of the border, more than 100 kilometres away from Nagorno-Karabakh. In late August, the Azerbaijani Ministry of Defense reported that the Armenians had shelled Agstafa district, near the town of Gazakh. In early September, Armenian troops allegedly fired on Azerbaijani positions even farther north, in Gazakh district.

The sides continued to accuse each other of using heavy weapons throughout the rest of the year. Azerbaijani forces were reported on 9 December to have fired more than 1,500 times on positions in Nagorno-Karabakh using tanks and grenade launchers. The next day, a spokesperson for the Azerbaijani Ministry of Defense reported that Armenian forces had shelled Azerbaijani villages near the Line of Contact.

Diplomatic activity

There was a flurry of diplomatic activity in early 2015, as Armenian and Azerbaijani officials held a series of high-level meetings facilitated by ambassadors Igor Popov, Pierre Andrieu and James Warlick – the Russian, French and US co-chairs of the Minsk Group, part of the Organisation for Security and

Co-operation in Europe (OSCE), respectively. On 27 January, the co-chairs met Azerbaijani Foreign Minister Elmar Mammadyarov in Krakow, where they expressed concern about incursions across the Line of Contact earlier that month and asked Baku to adhere to the ceasefire. On 5 February, they met Armenian Foreign Minister Edward Nalbandian in Munich, reiterating the importance of sharing data on missing persons through the International Committee of the Red Cross and calling for a resumption of the dialogue between the Armenian and Azerbaijani presidents.

During 16–19 February, the co-chairs met Azerbaijani and Armenian presidents, foreign ministers and defence ministers, as well as representatives of the Nagorno-Karabakh authorities. Armenian President Serzh Sargsyan and Azerbaijani President Ilham Aliyev agreed to consider proposals for strengthening the ceasefire put forward by the ambassadors. However, this active diplomacy did not appear to reduce the violence.

As tensions between the sides increased in summer and early autumn, the international community urged all parties to exercise restraint. In September, the co-chairs again met Nalbandian and Mammadyarov, both separately and together. They discussed 'the immediate need to reduce tensions along the Line of Contact and international border, to advance negotiations on a lasting settlement, and to implement confidence-building measures'. While the parties made no substantive agreements, the foreign ministers promised to continue preparations for a presidential summit scheduled for later in the year. The co-chairs called on Sargsyan and Aliyev to implement an OSCE mechanism for investigating ceasefire violations, stating that 'without such a mechanism, the sides will continue to blame each other for initiating deadly attacks on the Line of Contact and Armenia–Azerbaijan border'. By the end of the year, only Yerevan had agreed to discuss the specifics of this proposal.

In late October, the co-chairs conducted another three-day trip in which they met the Armenian and Azerbaijani presidents and foreign ministers, as well as representatives of the Nagorno-Karabakh authorities. As the ambassadors made a routine crossing of the Line of Contact, gunfire nearby forced monitors from the OSCE to take cover. The organisation described the incident as 'a deliberate effort to undermine progress toward a settlement'. Nonetheless, the summit between Sargsyan and Aliyev was held in Bern on 19 December. The presidents discussed recent violent incidents, as well as the use of heavy weaponry along the Line of Contact and the Armenia–Azerbaijan border. While no formal agreements were reached, the leaders reiterated their willingness to continue the talks.

Legal disputes

Despite persistent diplomatic efforts, the conflict had taken on legal dimensions earlier in the year. The unfolding court battles between the sides focused primarily on private citizens in Azerbaijan and Nagorno-Karabakh territory, as well as displaced persons. On 10 March, a court administered by the Nagorno-Karabakh authorities upheld a December 2014 decision in which two Azerbaijani men were found guilty of murder and espionage, as well as illegally crossing the border and possessing weapons.

On 16 June, the European Court of Human Rights handed down judgments on two of more than 1,000 cases related to the conflict. In both, one brought against Armenia and the other against Azerbaijan, the court ruled that there had been violations of the right to the peaceful enjoyment of property, the right to respect for private and family life, and the right to an effective remedy under the European Convention on Human Rights (Articles 1, 8 and 13 respectively). The court held that the peace process led by the OSCE did not nullify Baku and Yerevan's responsibility to protect civilians. It recommended that both sides establish effective, accessible mechanisms for displaced persons to make property claims.

On 28 December, a court in Baku sentenced Azerbaijani journalist Rauf Mirkadirov to six years in prison for spying for Armenia. His lawyer claimed that the charges were politically motivated, a view that was supported by Human Rights Watch and Amnesty International.

The risk of an all-out confrontation between Armenia and Azerbaijan remained low. Yet, due to the increasingly volatile situation on the ground and the lack of real progress in reaching a mutually acceptable compromise and building trust, the prospects for a long-term settlement between the parties remained as remote in 2015 as they were the previous year.

Central Asia

Key statistics	2014	2015
Conflict intensity:	Low	Low
Fatalities:	30	60
New IDPs:		
New refugees:	1,000	

In 2015, Central Asian governments faced threats from regional and international extremist groups, resulting in numerous arrests and security operations, as well as an expanding crackdown on citizens' access to the media and religious freedoms.

The governments of Central Asia arrested and tried alleged terrorists throughout 2015. Large numbers of citizens faced charges for belonging to banned organisations such as the Islamic Movement of Uzbekistan, Tablighi Jamaat, Hizb ut-Tahrir, Group 24 and Jamaat Ansarullah. Kyrgyzstan's authorities detained alleged members of Hizb ut-Tahrir across the country, and claimed to have arrested several of the group's leaders. The security services of Kazakhstan and Kyrgyzstan also apprehended and tried many people suspected of planning to travel to the Middle East to fight alongside jihadist organisations such as the Islamic State (also known as ISIS or ISIL), as well as those who had allegedly returned from the region to carry out terrorist attacks at home.

On several occasions, Tajikistan jailed and imposed large fines on members of the banned Salafist sect. In Kazakhstan, a resident of Atyrau was sentenced on 27 April to three years in prison for 'taking part in military conflicts abroad', after he was found guilty of fighting alongside Russian-backed separatists in eastern Ukraine. In May, a court in Osh designated Jabhat al-Nusra, Imam

Bukhari Jamaat and Jannat Oshiqlari as terrorist organisations, thereby banning all activities by the groups.

Meanwhile, the Islamic Movement of Uzbekistan (IMU) pledged allegiance to ISIS in August. The IMU, most of whose members had been based in Pakistan since 2001, had conducted terrorist attacks across the region for more than two decades. The group's main goal was to establish a Central Asian Islamic state subject to sharia law. Yet it was unclear whether the alliance with ISIS would increase the material support it received.

In September, a group led by Abduhalim Nazarzoda, Tajikistan's deputy defence minister, launched an attack on a police station near Dushanbe, killing 26 people. The Prosecutor-General's Office charged him with treason, terrorism, sabotage and creating an extremist group. They also claimed that Nazarzoda had ties with ISIS, although he had not publicly sworn allegiance to the group. The security forces tracked the attackers to Romit Gorge, east of Dushanbe, where they engaged in a battle that led to the deaths of at least 45 people, including Nazarzoda, ten of his supporters and 13 soldiers. In early October, several more residents of Dushanbe were detained in connection with the attacks, and, in mid-November, two more members of Nazarzoda's group were killed during an operation in Romit Gorge.

Following Nazarzoda's revolt, the authorities intensified their crackdown on the Islamic Renaissance Party of Tajikistan (IRPT), which the Ministry of Justice had banned in August, claiming that the organisation was too small to qualify as an official party. The IRPT had been the only officially recognised Islamist political party in the former Soviet Union, and was one of the signatories to the 1997 General Agreement on the Establishment of Peace and National Accord that ended Tajikistan's civil war. The authorities accused the IRPT leadership of involvement in Nazarzoda's attack, and subsequently detained its first deputy chairman – albeit without explaining the reason for his arrest. By the end of September, the IRPT had been added to the government's list of terrorist groups.

The suppression of the IRPT continued as the year drew to a close. Prosecutors announced in October that, since the beginning of the crackdown, 23 of the organisation's officials had been arrested on charges ranging from terrorism and inciting religious hatred to fraud and other economic crimes. Moreover, the government asked all foreign media accredited in Tajikistan to refrain from disseminating information related to, or generated by, the IRPT. The UN Office of the High Commissioner for Human Rights voiced concern about the situation, calling on Tajikistan's government to ensure that its anti-extremism measures complied with international human-rights law.

The threat to human rights was evident throughout Central Asia in the year, as governments imposed restrictions on the lesbian, gay, bisexual and transgender (LGBT) community, as well as on media and religious freedoms. In Kyrgyzstan, there was also an increase in violence against that community. In one instance, on 17 May, a group of activists attacked a party attended by members and supporters of Bishkek's LGBT community; the authorities interceded before anyone was injured, and several of the attackers were arrested on charges of hooliganism.

Meanwhile, Tajikistan's government further tightened its control of citizens' access to the media. On 19 June, law-enforcement officials blocked Facebook, YouTube and the Russian-language social media site Odnoklassniki, following the publication of a video depicting a former police commander who had allegedly defected to ISIS. In November, the country's parliament passed an amendment to the Law on Combating Terrorism, which a spokesman for the lower house described as providing the State Committee for National Security with the 'right to restrict access to electronic media in areas where counter-terrorism operations are being carried out'. As a consequence, there were concerns that the government had acquired a legal basis for shutting down social-media sites and even the internet as a whole.

Tashkent and Dushanbe were particularly aggressive in suppressing religious freedoms. On 13 April, Tajikistan's government banned citizens under the age of 35 from participating in the hajj. Its Committee for Religious and Cultural Issues declared that the move was designed to give older Muslims the opportunity to make the pilgrimage, in light of the fact that Saudi Arabia would only allow a set number of people from each country to visit Mecca. In September, Uzbekistan's education ministry banned people under the age of 18 from attending Friday prayer services and Islamic holiday celebrations. Under the new measure, the parents of those who broke the ban faced a fine of more than 2,000,000 som (US$750). The government justified the move by claiming that religious activities were distracting children from their studies.

Central Asian governments attempted to manage the threat of extremism, from at home and abroad, in 2015. However, the resulting policies provided cover for the states in the region to crackdown on civilians' rights or were used as an excuse to undermine opposition parties. In 2016, these efforts are likely to continue, or even intensify, especially as neighbouring Afghanistan becomes more destabilised.

Russia (North Caucasus)

Key statistics	2014	2015
Conflict intensity:	Medium	Medium
Fatalities:	300	200
New IDPs:		
New refugees:		

Despite carrying out in December 2014 one of the most violent and well-orchestrated terrorist attacks in recent memory, the Caucasus Emirate (CE) – the militant jihadist group that aims to establish an Islamic caliphate in the North Caucasus – was unable to sustain its momentum the following year. Moreover, two successive leaders of the group were killed in just a few months.

While Dagestan remained the most violent region of the North Caucasus throughout the year, the fighting there was much less intense compared to 2014.

Security personnel conducted a series of operations against groups of militants in Dagestan, leading to the deaths of many local cell leaders. During one raid, carried out in late March in the city of Khasavyurt, law-enforcement officers killed the leader of the terrorist group Kizilyar. The deadliest incident occurred in the Kizilyurt district of Dagestan, where an operation led to the deaths of nine suspected militants and two security officers, while two other law-enforcement officials were injured. In June, security officials conducted a raid in Khasavyurt, killing the leader of the Aukhovsky group. In early November, security officers conducted a special operation in Makhachkala region, killing the leader of the Makhachkala group; later that month, the authorities killed the leaders of the Tabaransky and Suleiman-Stalsky groups.

More importantly, the CE continued to lose its top leaders. Aliaskhab Kebekov, the CE's emir, was killed along with four other suspected militants during a raid in Dagestan in April. Shortly after his death, Magomed Suleymanov, also known as Abu Usman Gimrinsky, was reportedly chosen as the new leader of the group. Suleymanov had previously fought as a commander of the group's mountain sector in Dagestan and, like his predecessor, was a *qadi* (sharia judge) and an ethnic Avar from the region. Given their similar backgrounds, Suleymanov might have continued Kebekov's policies – but he had little opportunity to do so. Suleymanov and three other militants were killed on 11 August, during a counter-terrorism operation in the Untsukulsky district of Dagestan. Having lost its third emir in two years, the CE appeared to be growing increasingly weak.

The rise of the Islamic State

Suleymanov's death exacerbated the serious challenges faced by the CE, which included defections by its members to groups fighting in Syria and Iraq, and to Russian-based affiliates of the Islamic State (also known as ISIS or ISIL).

As the CE continued to lose ground in the North Caucasus, ISIS sought to fill the vacuum. Meanwhile, the authorities grappled with the interconnected issues of dealing with would-be insurgents travelling between Russia and the Middle East, and addressing the threat from ISIS-linked groups at home. Perhaps the most pressing issue for Kebekov and Suleymanov during their short tenures was managing the widening internal rift within the CE, caused by the growing influence of ISIS. On 13 June 2015, a commander in Chechnya pledged his allegiance to ISIS, arguing that he spoke for all militants in the republic. One week later, a group of leaders from across the North Caucasus claimed that the CE had pledged allegiance to ISIS. On 23 June, an ISIS spokesman announced the creation of an official *wilayat* (province) of ISIS in the region, called Wilayat al-Kavkaz.

Officials in Chechnya and Kabardino-Balkaria reported several instances of citizens attempting to travel to the Middle East to fight on the side of Islamic militants. On 15 April, the authorities began an investigation into two teenagers who were allegedly planning to travel to Syria for this purpose. Five days later, officers arrested a man in Nalchik, Kabardino-Balkaria's capital, as he attempted to travel abroad. Similarly, in August, several residents of the city were detained at the Russia–Georgia border in Kabardino-Balkaria as they attempted to leave for Syria. And, in September, officers in Chechnya's Shelkovsky district arrested six locals who were planning to travel to the Middle East, while a suspected ISIS recruiter was detained in Vedeno district.

Wilayat al-Kavkaz released in early October a video calling for all jihadists in the region to focus on the fight against Russia's security services instead of travelling to Syria to fight there. In the recording, one of the group's leaders stated that militants should 'obey the order of the Caliph, the Qadi of the caliphate, and participate – I mean the Muslims of the Caucasus, that they [should] go out and wage jihad in the Caucasus'.

This appeared to have an effect, as the region subsequently experienced a marked increase in incidents related to ISIS. In the deadliest clashes, a total of 14 alleged ISIS recruiters died during two counter-terrorism operations in Nalchik. Yet websites linked to other extremist groups in the region questioned the militants' connection to ISIS. Elsewhere in the North Caucasus, alleged members of ISIS were killed in multiple incidents in Dagestan, Kabardino-Balkaria and Chechnya.

ISIS affiliates have not demonstrated any more capacity in their efforts to undermine intense Russian government counter-terrorist efforts than other militant organisations.

Muslim civilians in danger

Throughout the year, both law-enforcement officers and militants targeted Muslim civilians who appeared devout. In late January and early February, several groups of men who had beards and wore traditional clothing were detained in Grozny. In one of these incidents, two men were allegedly beaten and forced to shave off their beards. On 3 March, the authorities arrested 30 worshippers in Dagestan's Kirov district, without providing an official explanation. In early May, between ten and 15 worshippers at a mosque in Buinaksk were remanded in custody, before being released later in the day; a week later, worshippers at another mosque in Buinaksk were detained for an identity check and released the next day; and the deputy imam of a mosque in Kurban was arrested in late May on charges of possessing illegal weapons (a court in Makhachkala extended his detention by one month on 1 June). In two incidents in Buinaksk in November, the security services detained a total of 50 worshippers, who were eventually released. Meanwhile, residents of Grozny and other large towns reported in October that dozens of men with beards had been arrested.

In September, an imam was killed in Khasavyurt district, and a family of three in Derbent. The latter incident was reportedly spurred by the fact that one member of the family was a fortune teller. Throughout 2015, the CE regularly targeted civilians engaging in practices banned by the Koran. While Kebekov forbade CE fighters from targeting civilians, that order was apparently annulled after his death. Militants in the North Caucasus continued to target places of worship and Islamic education centres. Police officers defused two improvised explosive devices discovered at an Islamic school in Derbent on 2 November, detaining four suspects in the process. And, on 1 December, militants gunned down an imam in Khasavyurt.

Yet there was at least once positive development in the region, with the continuing success of the campaign to clear mines planted during the Chechen conflicts of the 1990s and early 2000s. As a consequence, by September, sappers had removed mines from 2,000 hectares of land in Achkhoi-Martan and Itum-Kalinsky districts.

Developments in 2015 suggested that the CE may be waning, especially as it faced ongoing leadership issues and defections. While ISIS may be attempting to fill the void by expanding its operations in the region, it too faced difficulties, as the Russian government continued to crack down on extremist groups and citizens alike.

Ukraine

Key statistics	2014	2015
Conflict intensity:	High	High
Fatalities:		9,000*
New IDPs:	650,000	
New refugees:	250,000	

*Estimated conflict deaths since the start of the conflict

In 2015, Ukraine continued to face an armed insurgency in Donetsk and Luhansk regions. While the parties agreed on a new peace deal, it proved difficult to fully implement. According to the Special Monitoring Mission (SMM) of the Organisation for Security and Co-operation in Europe (OSCE), the Ukrainian military and Russian-backed separatists continually violated ceasefires throughout 2015. Fighting between the sides peaked early in the year with the battle for the town of Debaltseve, in Donetsk region, before declining dramatically in September, after they implemented a new ceasefire.

In January and early February, the separatists expanded the areas of eastern Ukraine under their control by more than 300 square kilometres. The rebels advanced on the outskirts of Donetsk city, pushing both northwest to take the airport and southwest towards the village of Buhas. On 22 January, as the opposing forces made small territorial gains while sustaining heavy casualties, the separatists captured Donetsk airport, prompting the military to conduct a chaotic retreat in which the bodies of fallen soldiers were left behind. In Luhansk region, the rebels made small gains along their northern flank. They also shelled Mariupol and Kramatorsk, cities far inside government-controlled territory.

Fluctuating violence, flawed agreements

In a bid to halt the escalating violence, German Chancellor Angela Merkel and French President François Hollande travelled to Kiev and Moscow during 5–6 February, meeting Ukrainian President Petro Poroshenko and Russian President Vladimir Putin to discuss a new ceasefire agreement. The leaders planned to meet in Minsk on 11 February to finalise the document, working in parallel with talks between the Trilateral Contact Group (made up of negotiators from Ukraine, Russia and the OSCE) and separatist representatives.

On 12 February, after more than 16 hours of negotiations, the leaders endorsed a set of 'implementing measures' intended to revive the principles of the original Minsk Protocol (agreed in September 2014). The new document, known as 'Minsk II', stipulated that: a ceasefire begin at midnight on 15 February; both sides remove their heavy weapons from the battlefield in the first two weeks of the ceasefire; the SMM carry out enhanced monitoring of the first two provisions using all technical means necessary; local elections be conducted in separatist-controlled areas after the withdrawal (in accordance with agreed conditions); all combatants be given amnesty, and all hostages released; and humanitarian-aid agencies be given free access to civilians in need. Minsk II also stated that Ukraine would make constitutional reforms by the end of the year, devolving authority to areas held by the separatists (the details of which would be set after consultation with their representatives). Furthermore, after the constitutional reforms were implemented, Russian forces would relinquish control of Ukraine's border with Russia; all foreign armed groups, military equipment and mercenaries would withdraw from Ukraine; and several working groups would be created to hash out the details of implementation, under the aegis of the Trilateral Contact Group. The agreement included withdrawal distances for various weapons systems: at least 50km for artillery with a calibre of 100mm or more; 70km for most multiple-launch rocket systems; and 140km for *Tornado-S*, *Uragan* and *Smerch* multiple-launch rocket systems, as well as *Tochka* and *Tochka-U* tactical missile systems. The Ukrainian military would withdraw these weapons from the

Line of Contact as of 12 February, while the separatists were to withdraw from the September 2014 Line of Contact.

Minsk II placed Ukrainian domestic political compromise at the centre of the negotiation process. It was more precise than the September 2014 agreement in sequencing significant provisions, specifying that Russia would not have to cede control over the Russia–Ukraine border until Ukraine reformed its constitution. In this respect, the agreement required major concessions by Kiev. But the document had significant flaws, likely due to the frantic negotiation process. Some provisions were vague, and therefore open to interpretation. Crucially, it failed to specify an armistice line – a flaw that immediately became apparent in the intensifying battle for Debaltseve.

In the lead-up to the Minsk II negotiations, pro-government forces had controlled the town but were surrounded on three sides, as the separatists continued to gain ground. The main highway leading to government-held territory had been almost completely destroyed by artillery fire, and thoroughly mined. By 17 February, separatist forces benefiting from significant Russian support had fought their way into the town proper. The Ukrainian military was forced into another haphazard retreat in which it suffered heavy losses. Estimates of the casualties varied, but at least 100 Ukrainian soldiers were captured. After the capture of Debaltseve and surrounding territory, the ceasefire largely held, with the Line of Contact a rough semicircle that ran from the Ukraine–Russia border to north of Luhansk city, northwest of Gorlovka and Donetsk city, and west of Novoazovsk, on the Sea of Azov.

By the end of March, Minsk II's shortcomings had become fully apparent. While some heavy weapons were withdrawn, the sides only partially implemented some key provisions of the agreement. Due to its limited capabilities, the SMM was unable to verify the tiered withdrawal requirements. While Ukraine's parliament passed some of the laws required by Minsk II, aspects of the legislation were not consistent with the deal, and there was no progress in the political process.

There were also battles in several areas along the Line of Contact. Between April and June, the SMM reported a gradual increase in the tempo of fighting in separatist-held territory, especially near Donetsk airport and Shyrokyne, a village 20km east of Mariupol. On 13 April, intensifying violence forced OSCE monitors to withdraw from their post in Shyrokyne. For several weeks afterwards, there were reports of large-scale clashes and troop movements near the village.

The fragile ceasefire was threatened again on 3 June, when brief but intense battles broke out near Mariinka, a town on a strategically important highway near checkpoints along the Line of Contact. The SMM reported the movement of tanks and heavy weapons in the area, and both sides' use of artillery and *Grad* multiple-launch rocket systems. Throughout the month, monitors also witnessed fighting around Donetsk airport, some of which involved the use of mortars and heavy artillery.

On 1 September, following a series of reports that the separatists were being reinforced with heavy weapons, the parties agreed to a new ceasefire. This time, almost all fighting ceased – albeit with the SMM continuing to report movements of military hardware in separatist-held areas, as well as sporadic clashes involving small arms. There was limited progress in removing weapons from the conflict zone, as the sides began to implement a 13 April agreement that called for the withdrawal of all types of tanks and artillery, as well as heavy weapons with a calibre of more than 100mm. And, on 29 September, the Trilateral Contact Group reached an agreement on moving weapons with calibres of 100mm or less at least 15km away from the Line of Contact.

Poroshenko, Putin, Merkel and Hollande met in Paris on 2 October to discuss the conflict and the implementation of Minsk II. The talks focused on four issues: security, including the withdrawal of heavy weapons and the safety of OSCE monitors; the political process, particularly local elections in rebel-held territory; the humanitarian situation; and economic issues, including Russian gas supplies to Ukraine. The key aim of the meeting was to agree on a new timeline for the local elections. Putin agreed to pressure the rebels to call off their own vote, scheduled for 18 October and 1 November. Poroshenko agreed to hold official elections in rebel-held areas in spring 2016.

Meanwhile, both the Ukrainian military and the separatists continued to withdraw weapons from the Line of Contact. On 17 October, the SMM received written declarations from both sides that they had removed mortars with calibres of up to 120mm. On 28 October, they announced that they had withdrawn all heavy weapons from the front-line near the Artemivsk sector of Donetsk region. However, the reports were difficult to verify, and the weapons had likely been withdrawn to locations from which they could be easily redeployed. During a briefing at the Joint Centre for Control and Coordination on 25 November, the Ukrainian government declared a 'silence regime', under which the military would not return fire if attacked by separatists in the conflict zone.

Yet, despite these positive moves, the sides failed to implement key provisions of Minsk II, including those related to developments on the ground and legislation in Kiev. Moreover, they continued to regularly violate the ceasefire, albeit on a far smaller scale than they had before September. Poroshenko, Putin, Merkel and Hollande discussed these issues again by phone on 30 December, agreeing to extend the timeline for implementing Minsk II into 2016. Following the talks, the French government stated that 'Angela Merkel, François Hollande, Vladimir Putin and Petro Poroshenko have reaffirmed their commitment to a ceasefire in eastern Ukraine and to a concerted pullout without delay of heavy weapons.'

Monitoring the conflict

In early March, the SMM reported that it was 'restrained in fulfilling its monitoring functions by security considerations including the lack of information on the whereabouts of landmines, and restrictions by third parties'. The organisation continued to declare that its activities were being limited by both government and separatist forces, which often prevented monitors from accessing areas on a daily basis, stopped them at checkpoints and detained them for short periods.

On 27 July, one monitor was injured when two SMM observation points in Shyrokyne came under fire. The day before, members of an SMM patrol narrowly escaped injury after being caught in a mortar and grenade attack while trying to cross into government-controlled territory from separatist-controlled areas near the town of Shchastya, 20km north of Luhansk city. On 9 August, four OSCE vehicles in Donetsk city were set alight, although no one was injured. In December, separatist fighters threatened an SMM patrol in Yurivka, a village in Luhansk, and security concerns relating to rebel movements subsequently prevented SMM monitors from reaching Kominternove, a village northeast of Mariupol, just outside government-controlled territory. A few days after the monitors were able to enter Kominternove, a nearby SMM patrol vehicle was damaged by rounds from small arms.

The rise of nationalist groups

The rise of violent nationalist groups in other regions of Ukraine was an unintended result of the conflict in the east. A spate of incidents involving these organisations demonstrated the difficulty Kiev faced in managing radical actors empowered by the popular uprising against the previous government. Ukrainian law-enforcement officers clashed on 11 July with members of nationalist group Right Sector in Mukachevo, in the Zakarpattia region of southwestern Ukraine. Around 20 Right Sector militants engaged in a protracted gun battle with the security forces in which at least three people were killed and 13 others wounded. The fighting had broken out due to a dispute between the group and local authorities over revenues from a cigarette-smuggling operation.

After a constitutional-reform bill passed its first reading in parliament in late August, far-right party Svoboda staged a protest that quickly became violent. A grenade was thrown at national guardsmen protecting the parliament building, leading to a clash in which four guardsmen were killed and more than 100 protesters and police wounded. Around 30 people were arrested over the incident.

Disrupted civilian life

The Office of the UN High Commissioner for Human Rights reported that, by mid-November, around 9,000 people had been killed and at least 20,700 injured in the conflict. A constant in the war was its devastating effect on civilians: the International Organization for Migration found that 1.6 million people had been internally displaced by the fighting by the end of the year, while the Office of the UN High Commissioner for Refugees stated that, by the end of the year, more than 1m refugees had fled to neighbouring countries, particularly Russia (a figure that did not include unregistered migrants).

The fighting around Debaltseve in February was particularly damaging to the civilians who remained there, most of whom were elderly people. The United Nations reported that, after the fighting ceased, 500 civilians were found dead in their homes. Amnesty International described the situation in the area as 'catastrophic', stating that there was a lack of basic necessities among people attempting to shelter from heavy shelling.

The organisation published on 22 May a report that accused both sides in the conflict of abusing prisoners of war, some of whom stated that they had been 'beaten until their bones broke, tortured with electric shocks, kicked, stabbed, hung from the ceiling, deprived of sleep for days, threatened with death, denied urgent medical care and subjected to mock executions'. Kiev subsequently issued a statement denying all allegations in the report. Zeid Ra'ad al-Hussein, UN high commissioner for human rights, noted in July that 'even with the decrease in hostilities, civilians continue to be killed and wounded'. The UN monitoring mission in Ukraine argued that, in separatist-held areas, there were 'serious human rights abuses, intimidation and harassment of the local population perpetrated by the armed groups', and that it had 'received new allegations of killings, torture and ill-treatment, as well as cases of illegal deprivation of liberty, forced labour, looting, ransom demands and extortion of money on the territories controlled by the armed groups'.

The OSCE reported that, by July, up to 1.3m people in conflict-affected areas faced 'a serious water crisis due to damaged or destroyed water lines and acute water shortages'. Such infrastructure had been both targeted and accidently damaged by the fighting. The organisation also observed that the destruction of water facilities could have 'lasting consequences on food production as well as ...

heating systems'. Furthermore, the SMM released in late December a report cataloguing the threats to civilians in both government- and separatist-controlled areas of Donetsk and Luhansk regions. One of these related to the emergence of so-called 'parallel justice' systems in rebel-controlled areas, which were heavily criticised for their lack of transparency, professionalism and functionality.

Given the ebb and flow of the conflict, the potential for violence simmered just below the surface; it was likely that, until all parties demonstrated the necessary political will, Minsk II would remain unfulfilled and the conflict would continue.

Chapter Eight

Latin America

Central America (Northern Triangle)

Key statistics	2014	2015
Conflict intensity:	High	High
Fatalities:	15,000	17,000
New IDPs:		
New refugees:	3,000	

Throughout 2015, corruption and organised crime caused severe security problems and political instability in Central America's 'northern triangle', comprising El Salvador, Guatemala and Honduras. The resignation and imprisonment of Guatemalan President Otto Pérez Molina, who was implicated in a corruption scandal, caused a shift in his country's political environment. Meanwhile, criminal violence assumed troubling new forms in El Salvador, where the gangs intensified their coordinated armed campaign against the government.

As a consequence, the country recorded 6,670 homicides in the year, or more than 18 per day. This was a rise of 70% from 2014, and the largest increase in El Salvador's homicide rate in recent history. There were also 106 registered cases in which three or more people were killed in 2015, more than had taken place there in the preceding three years.

In January 2015, the most prominent leaders of El Salvador's two dominant *maras* (territorial gangs formed predominantly of young men) were returned to

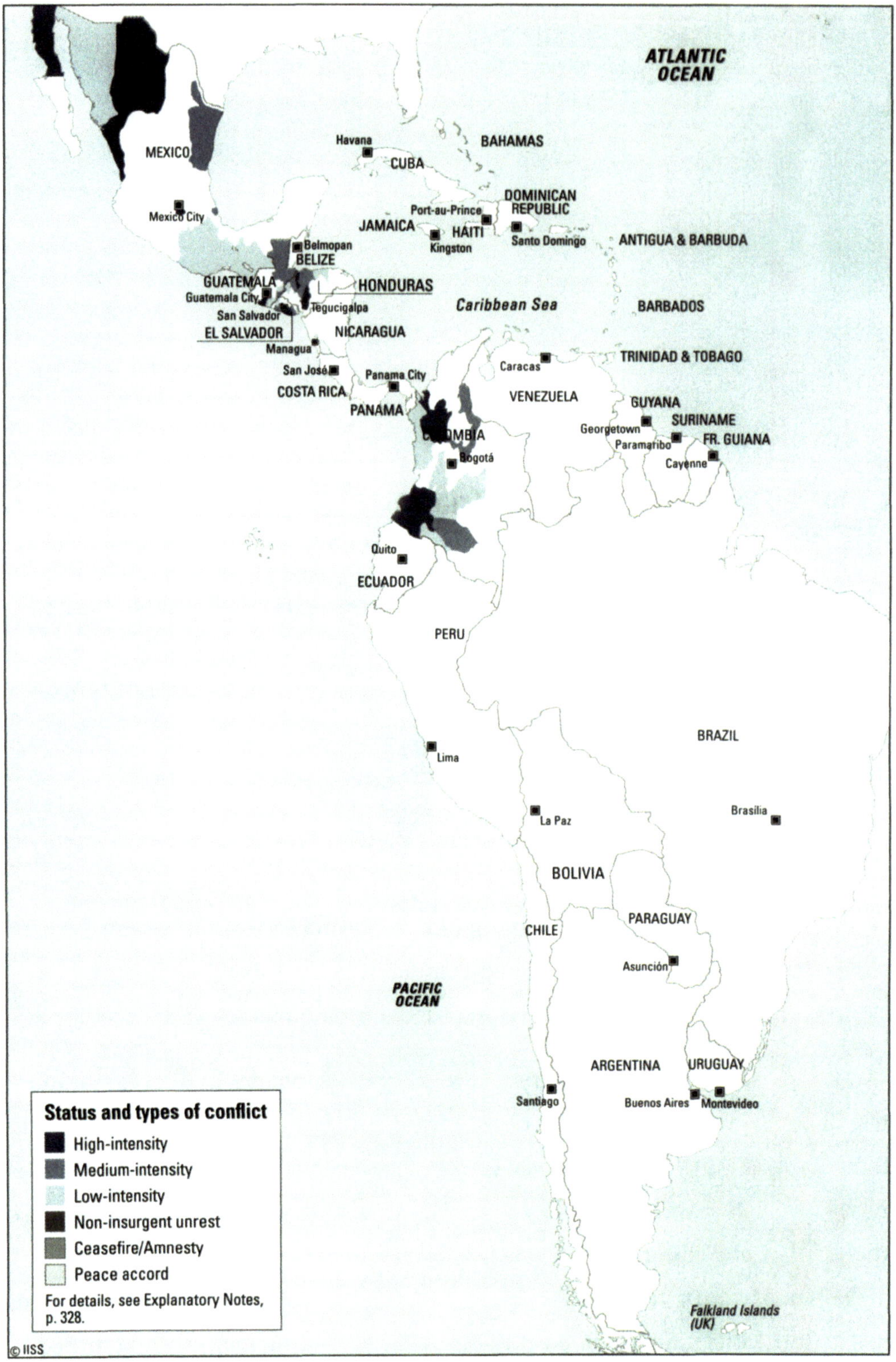

ATLANTIC OCEAN
BAHAMAS
Havana
CUBA
MEXICO
Mexico City
DOMINICAN REPUBLIC
Port-au-Prince
HAITI
Santo Domingo
JAMAICA
Kingston
ANTIGUA & BARBUDA
Belmopan
BELIZE
GUATEMALA
Guatemala City
HONDURAS
Caribbean Sea
San Salvador
Tegucigalpa
BARBADOS
EL SALVADOR
NICARAGUA
Managua
Caracas
TRINIDAD & TOBAGO
San José
Panama City
COSTA RICA
VENEZUELA
GUYANA
PANAMA
Georgetown
SURINAME
COLOMBIA
Paramaribo
FR. GUIANA
Bogotá
Cayenne
Quito
ECUADOR
PERU
BRAZIL
Lima
La Paz
Brasília
BOLIVIA
PARAGUAY
CHILE
Asunción
PACIFIC OCEAN
ARGENTINA
URUGUAY
Santiago
Buenos Aires
Montevideo
Status and types of conflict
High-intensity
Medium-intensity
Low-intensity
Non-insurgent unrest
Ceasefire/Amnesty
Peace accord
For details, see Explanatory Notes, p. 328.
Falkland Islands (UK)
© IISS

Zacatecoluca prison, a high-security facility. A 2012 truce between the gangs and the government stipulated that the leaders be transferred to low-security jails, allowing them greater freedom of movement and improving their ability to communicate with the outside world.

But the agreement collapsed due to rising violence – reportedly including the use of improvised explosive devices – linked to the *maras'* broad strategy of forcing the government to implement policy changes. The gangs' leaders subsequently sent open letters to the government requesting that civil-society mediators be permitted to engage in a new truce negotiation. The unprecedented wave of attacks against police officers and even army personnel came in direct response to the government's toughened security strategy, which had led to military confrontations with the gangs. Police intelligence published in April warned that the groups were planning to increase the frequency and intensity of their attacks against police officers, and to step up their combat training 'in order to increase pressure on the government' by maintaining a high homicide rate. Mara Salvatrucha (also known as MS13) was thought to be particularly committed to this approach.

The focus on government targets resulted in high casualties among police officers, a record 61 of whom were killed in armed attacks by gangs during 2015. This was an increase of 56% from the previous year. Furthermore, 24 members of the armed forces were killed in gang-related incidents – more than double that in 2014. The National Police reported in September 2015 that 423 gun fights between officers and alleged gang members had occurred since the beginning of the year, an annual increase of 171%. These battles resulted in the deaths of 197 suspects.

The most disruptive show of force from criminal groups came during 27–30 July, when normal life in San Salvador was suspended by a gang-enforced strike among public-transportation workers. The order to shut down the city's transportation network, affecting around 1,000 buses and minibuses, reportedly came from the Revolucionários faction of Barrio 18. Nine workers were killed by the gangs for violating the order to strike, seven of them on its first day.

According to Roberto Soriano, president of the Public Passenger Transportation Association, the move cost the transportation sector an estimated US$3.2 million and the cities' businesses around US$60m. The government announced on 28 July that it had arranged for 120 vehicles, including buses and trucks, to provide transportation in the capital. Two days later, army troops were patrolling San Salvador and its periphery, some in Humvees. The attorney general's office issued arrest warrants for 300 gang members thought to be involved in the strike. In August, President Salvador Sánchez Cerén announced that the police and military reinforcements sent to protect public transportation during the shutdown would be permanently deployed on the city's streets.

Amid national outcry against the disruption caused by the *maras*' growing boldness, the Salvadoran Supreme Court designated Mara Salvatrucha and Barrio 18 as terrorist organisations. The measure allowed the courts to apply the Special Law against Acts of Terrorism to the gangs' estimated 60,000 members, as well as their supporters. Under the law, those convicted of joining a gang could be sentenced to a maximum of 60 years in prison (in contrast, homicide and aggravated homicide carried maximum sentences of 20 years and 50 years respectively). The court also ruled that it was constitutional to tap the phones and freeze the bank accounts of third parties with ties to the gangs. These changes followed similar developments in Honduras, where congress approved changes to the penal code that would increase jail time for gang members convicted of any crime from 3–6 years to 20–30 years (albeit without designating them as 'terrorists'), and to allow gang leaders to be jailed for a maximum of 50 years.

El Salvador's army gained a formal role in internal security, with its establishment of three special-forces battalions specialising in anti-crime operations. Comprising 200 troops each, these battalions began training in April, in preparation for their deployment to the most violent areas of the country. The despatch of a 60-member group on 21 May marked the first time that army special forces had been deployed since 1992, when the Salvadoran civil war came to an end. The special-forces groups had been disbanded after the conflict due to concerns about their involvement in human-rights abuses while on law-enforcement duty.

The organisational structure of the new battalions would reportedly mirror that of the war-era Immediate Reaction Infantry Battalions, an army special-forces group responsible for combating left-wing guerrilla organisations (one of which had been led by Sánchez Cerén).

Militarisation and security challenges

The Honduran security services also expanded in the year, as the Military Police of Public Order (PMOP) took on new roles and a greater number of personnel. In early 2015, President Juan Orlando Hernández put a bill before congress stipulating that the force be given a permanent status in the constitution, only for the proposal to be rejected – with opposition lawmakers claiming it would establish a militarised law-enforcement structure that threatened democracy. They also argued that more should be done to strengthen Honduras's under-resourced civilian police agencies. Nonetheless, Hernández responded by calling for a referendum on the issue, and the vote was scheduled for 2017. The president subsequently announced that the PMOP's standing force comprised 3,228 personnel spread across six battalions, more than two-thirds of whom had joined in the two years since it was established.

Guatemala's National Defence Ministry approved in February the creation of a Special Reserve Corps for Citizen Security. Organised into eight squadrons, the corps was designed to support the law-enforcement activities of other military units, providing security in several cities and towns.

The new Maya–Chortí Task Force announced the same month by Honduras and Guatemala, set up to support a planned customs union between the countries, also combined military activities with policing. Forming part of an anti-crime border-security mechanism that Hernández called a 'land shield', the force began operations in March with 600 personnel drawn from the Honduran and Guatemalan police and armed forces, as well as intelligence, customs and migration agencies. However, Maya–Chortí covered only 15% of the porous 256-kilometre border between Honduras and Guatemala. The border area continued to be troubled by violence, with civilians forced to follow rules enforced by criminal groups – or at least acquiesce to illicit trafficking in their neighbourhoods.

Although the weak institutions of the northern triangle struggled to reduce the influence of gangs in suburbia and isolated rural areas, they received a significant boost in their efforts to address the underlying causes of criminal violence in late 2015, when the US Congress approved a US$750m aid package called the Alliance for Prosperity. It was hoped that, by strengthening a broad range of institutions in the region, the measure would reduce the problems that drove many civilians to travel illegally to the US. These 'interlocking security, governance and economic challenges', as US Vice President Joe Biden called them, included criminal violence, corruption, unemployment, lack of education and underinvestment. While the aid package potentially provided more than double the funding Washington usually allocated to all of Central America, three-quarters of the money would only be disbursed after the recipient countries took 'effective steps' to address the causes of violence and migration, and to improve transparency and fiscal discipline.

The resignation and arrest of Guatemalan President Otto Pérez Molina indicated the scale of the corruption problem in the northern triangle. The president was suspected of playing a major role in a bribery scheme involving Guatemala's customs system, and of having illegally collected around US$3.7m from companies importing products into the country in 2014 and 2015. These alleged activities were uncovered during a joint investigation by the attorney general's office and the UN-backed International Commission against Impunity in Guatemala. Pérez Molina resigned on 3 September, just three days before the general elections, and was arrested and imprisoned hours later. It appeared that he would stand trial with his former vice-president, Roxana Baldetti, who was thought to have helped lead the operation, and had been forced to resign in May.

The scandal had a major effect on Guatemalan politics: voters participated in the elections with the image of Pérez Molina's arrest fresh in their minds. As a result, former television comedian Jimmy Morales – who had little experience in politics – came first in the initial round, and went on to win the presidency comfortably in the run-off vote, held on 26 October. A member of the centre-right National Convergence Front, Morales won 67% of the vote, while centre-left candidate and former first lady Sandra Torres received just 33%.

Despite facing these considerable governance and security challenges, Honduras and Guatemala both benefited from falling homicide rates. The decline was particularly steep in Honduras, where there were 5,047 homicides in 2015, 14% fewer than in the previous year. Meanwhile, there were 5,718 homicides in Guatemala, a decrease of 3.4%. The governments in both countries attributed these trends to increased efforts to strengthen law-enforcement bodies.

Nonetheless, according to an intelligence report by El Salvador's National Police, Mara Salvatrucha was receiving combat training in Guatemala from Mexico's Los Zetas cartel. A relationship of this kind between two very different transnational gangs had the potential to upset the fragile balance of power between Mara Salvatrucha and Barrio 18. Another possible source of instability was the rift in the Salvadoran branch of Barrio 18, with the Revolucionários continuing to operate independently of the gang's leadership. An escalation of this infighting threatened to fracture Barrio 18 across the region, and to spark attacks from rival organisations seeking to capitalise on the group's divisions.

Colombia

Key statistics	2014	2015
Conflict intensity:	Medium	Medium
Fatalities:	500	350
New IDPs:	135,000	
New refugees:	1,750	

The armed conflict between the Colombian government and FARC underwent an unprecedented de-escalation in 2015, with a drastic reduction in casualties, as well as battles between the sides. The peace process continued to make progress as it entered its third year, albeit with the government receiving criticism from the opposition in congress for the concessions it made to the guerrillas. A resurgence of intense fighting between May and July indicated that there remained obstacles to the process, and to any post-conflict settlement. Yet, despite these challenges, the sides announced on 15 December that they had reached an agreement on the most controversial issue in the negotiations: transitional justice, including reparations for the victims of conflict.

Groundbreaking agreement

Transitional justice was the fourth of five points on the agenda for the negotiations, held in Havana, and an issue on which no previous Colombian administration had been able to forge a deal. Deals had previously been reached on rural reforms, political participation of former rebels and cooperation to tackle the illegal-drugs trade. Negotiations in early 2016 concentrated on steps for the disarmament of FARC fighters and on ending the conflict.

The agreement followed a turbulent year in which the sides came under growing pressure to make quicker progress in the talks – pressure heightened by sporadic armed clashes in the Colombian countryside. FARC had been adamant that its fighters not serve time in jail for their crimes, while the group's victims stated that it was unacceptable to grant the insurgents full amnesty (as had occurred in peace agreements in other countries). As a consequence, the accord was only reached after 17 months of negotiations, far longer than those for any other point on the agenda.

The sides agreed to fix the damage caused by the conflict through 'actions, political decisions and material contributions'. The government's chief negotiator, Humberto de la Calle, announced that FARC's contribution would be to assist in, inter alia, locating victims' bodies, clearing minefields and rebuilding infrastructure in areas affected by the conflict. Meanwhile, a special court would be established to hand down reduced sentences to combatants on both sides who confessed to war crimes or crimes against humanity. Those who confessed to their crimes before receiving a verdict would be confined in an unspecified location, though not in jail, for up to eight years. In contrast, those who did not confess would be sentenced to 15–20 years in prison, the harshest punishment under the agreement. Combatants who were not accused of war crimes or crimes against humanity would be given amnesty.

Where a military officer was charged with complicity in the crimes of his or her subordinates, prosecutors would have to prove that the accused had effective control of, or could have prevented, their criminal actions. The members of the military sent to prison under the special court would serve their sentences in military-detention centres. Furthermore, a truth commission would be established to aid the process of reconciliation after more than 50 years of war.

The agreement on transitional justice amplified the sense of accomplishment and optimism among Colombia's political elites and wider society. This sentiment had been evident at a September meeting between President Juan Manuel Santos and FARC leader Rodrigo 'Timochenko' Londoño, who shook hands for the first time and set 23 March 2016 as the deadline for completing the peace process. According to Gallup polls, the percentage of Colombians who thought the peace process was going well had risen from 38% in October to 52% shortly after the deal on transitional justice. In the same period, those who were optimistic about reaching a full peace agreement before the March deadline rose by 22 percentage points, to 45% (those who were pessimistic fell to 52%).

Nonetheless, despite these promising developments, any complete peace agreement would be put to a public referendum in 2016, allowing voters to have the final say on ending the war with FARC. According to Gallup, support for such a deal dropped from 71% in October to 52% in December. This may have been linked to the vocal opposition to the peace process from political sectors that remained popular with many Colombians – spearheaded by former president Alvaro Uribe.

Military escalation and de-escalation

The progress in the negotiations led to a sharp decline in military confrontations. The number of attacks carried out by guerrilla groups fell by 25% between

January and November 2015, in comparison to the same period the previous year. This was mainly due to FARC's unilateral ceasefire, which was implemented on 20 December 2014 and maintained for most of the following year. But there were frequent violations of the arrangement, with the group launching 21 attacks during the first half of 2015, according to Colombian think tank the Conflict Analysis Resource Center. The conflict quickly escalated after an ambush carried out on 15 April by FARC's Miller Perdomo column, which killed 11 soldiers in the department of Cauca. The fact that the ceasefire had been violated by such a devastating attack raised doubts about the ability of the FARC leadership to control its mid-level commanders. A spokesperson for the group expressed regret over the soldiers' deaths, but said the attack was motivated by the Colombian military's persistent harassment of the rebels.

Santos responded to the assault by ordering the military to resume its airstrikes against FARC, which had been suspended as a goodwill gesture. On 22 May, an aerial attack on one of the group's camps near the town of Guapi, in Cauca, killed 26 fighters attached to the 29th Front, the heaviest loss suffered by the rebels in three years. FARC subsequently called off its ceasefire, prompting Juan Pablo Rodriguez, head of the joint chiefs of staff, to order all troops to ready themselves for a rebel counter-attack. Colombia as a whole braced for a resurgence in guerrilla warfare, and it quickly materialised. Between 23 May and 25 May, the group carried out several attacks, most of them in its rural strongholds in the south and southwest. The departments of Narino, Caqueta and Antioquia were affected first and, by mid-June, 700 civilians in Cauca had been internally displaced. Offensives by guerrilla groups increased fourfold between May and July, while terrorist acts more than doubled between May and June, with 130 attacks in the latter month.

FARC's targeting of energy infrastructure exacerbated public outrage against the group. A series of bombings in Narino and Putumayo culminated in a 21 June attack on the Transandino oil pipeline, which caused 10,000 barrels of oil to spill into the Guisa river. The country's worst such incident in a decade, the spill stretched across 15 kilometres of river and, despite the authorities' efforts to contain it, reached the Pacific Ocean and polluted some beaches. After visiting the area, Minister of Environment and Sustainable Development Gabriel Vallejo López stated that 'those most affected are the least protected and most vulnerable classes and, of course, the fishermen in the area.' Aside from the damage caused to the environment and affected communities' sources of income, approximately 100,000 people in the city of Tumaco were left without a water supply for days.

Gallup reported in June that, for the first time, more Colombians were in favour of a military solution to the conflict than a peace deal (46% and 45% respectively). The following month, de la Calle urged FARC negotiators to increase their efforts to reach an agreement, arguing that the talks would break down if 'the Colombian people's patience runs out'. He added, 'it is possible that one of these days the FARC will not find us at the table in Havana.'

While it did not directly address the public hostility it faced, the group quickly attempted to ease tensions, announcing that a new unilateral ceasefire would begin on 20 July. Santos also appeared to sense that the peace process was in danger of collapsing, and announced the suspension of airstrikes against FARC camps that did not pose a 'threat to the population, the public forces or the infrastructure'. This time, FARC largely adhered to the ceasefire: the Conflict Analysis Resource Center reported that the intensity of the conflict reached an all-time low, with the group launching only six attacks between 20 July and 20 December. And Timochenko made a significant symbolic move on 1 October, ordering his subordinates to cease military training and instead pursue political and cultural studies.

Both sides used the term 'de-escalation' to describe their moves to gradually reduce the intensity of the conflict. This language allowed the government to avoid the political difficulties of declaring a full bilateral ceasefire before the end of the talks. With FARC repeatedly calling for a full truce, Santos declared on 28 October that his team had agreed to work with the group towards implementing a bilateral ceasefire, which would only begin once the peace talks had been concluded. Nonetheless, FARC's negotiators subsequently rejected some of Santos's conditions for a bilateral ceasefire, especially the stipulation that fighters be concentrated in pre-determined areas and monitored by military officials.

As well as massively reducing the frequency of its attacks, FARC cooperated with the government in clearing Colombia's minefields – a process that would 'take a generation, tens of thousands of specialists and lots of international aid' to complete, according to Minister of Defence Luis Carlos Villegas. But, despite the sharp decline in violence, the peace process led to an escalating war of words in Bogota. While the deal on transitional justice was praised by the Organisation of American States and the United Nations, other actors criticised the government for allegedly giving too much ground to the rebels. Former president Alvaro Uribe, a senator from right-wing party Democratic Centre and one of the most popular politicians in the country, was by far the fiercest critic of the process. Alfredo Rangel and Alejandro Ordóñez, two of Uribe's allies in government, respectively described the deal as 'disguised impunity' and 'an affront and a slap on the victims [because] it ignores their rights'. Human Rights Watch also condemned the agreement, stating that it would 'allow those most responsible for mass atrocities to completely avoid prison'.

In a tense congressional debate held in September, de la Calle clashed with Uribe and other members of the opposition. De la Calle responded to the accusation that the agreement was a 'comedy' of justice by reassuring lawmakers there would be no general amnesties, and that the provision of sentences other than imprisonment was necessary to complete the peace process. By December, Santos and his allies appeared to be winning the debate, and Uribe appeared increasingly isolated in his opposition to the peace process. Not only had there been an increase in public support for the effort, but the government's majority in congress ensured that it could pass bills, paving the way for a final agreement. When the government proposed on 14 December to put the issue to a popular referendum, only Democratic Centre voted against the motion.

Secondary threats

In contrast to the negotiations with FARC, the government's peace talks with the National Liberation Army (ELN) appeared to make little progress. As the second-largest insurgent group in both Colombia and the Western Hemisphere, the ELN was estimated by the country's military to have around 1,500 fighters. Given the decline in operations against FARC, many Colombians speculated that preliminary talks between Bogota and the ELN would be accelerated by the latter's knowledge that the military could devote more resources to other battles. Yet there was no evidence of this by the end of the year, more than two years after the talks had begun. Furthermore, it was reported on 28 December that the government of Venezuela (where many ELN leaders were based) had 'frozen' its support for the talks, due to deteriorating relations between Caracas and Bogota.

Several bold attacks by the ELN also cast doubt on the group's willingness to negotiate. On 27 October, it claimed responsibility for an assault that had killed 11 soldiers and one police officer in the department of Boyaca the previous day. The ELN also announced that it had taken two soldiers hostage. President Santos described the operation as an attack on democracy, due to the fact that those killed had been transporting voting cards from regional elections.

Earlier in the year, the security forces had killed part of the leadership of two influential ELN blocs: the Eastern Bloc and the Dario Ramirez Castro Bloc. The former was one of the insurgents' most militarily active groupings, while the latter was heavily involved in illegal mining in Bolivar and Antioquia departments. The ELN's attacks might have been intended to establish a position of strength ahead of a formal peace process. Alternatively, they could have resulted from a lack of coordination between the ideologically motivated central leadership and younger mid-level commanders more focused on criminal activities.

The economic motivations of some ELN fighters spoke to the significant risks posed by organised crime in a post-conflict Colombia. Both FARC and the ELN had wide experience of drug trafficking and illegal mining. Even more worrying, the country had many times been involved in unsuccessful disarmament and demobilisation processes. Until his death in 2015, the most-wanted drug trafficker in Colombia had been Victor Ramón 'Megateo' Navarro Serrano, leader of the only remaining faction of guerrilla organisation the Popular Liberation Army, which demobilised in 1991.

Furthermore, Los Urabeños and Los Rastrojos, Colombia's two largest crime gangs, developed out of factions of right-wing paramilitary organisations that demobilised in the 2000s. Both gangs had international connections and access to small arms, although the security forces engaged in fewer battles with them than with the country's guerrillas. The deputy head of the Colombia office of the UN High Commissioner for Human Rights stated in March that the gangs were most often responsible for violence against civilians perpetrated by non-state actors. He cited a series of gruesome killings carried out by Los Urabeños in the Pacific port city of Buenaventura. The group had emerged as the most aggressive gang in the country, frequently clashing with rivals such as FARC and the ELN. In October, a war between Los Urabeños and two guerrilla groups for control of drug-trafficking routes led the authorities in the department of Choco to declare a state of emergency. The latter described the conflict as having created a 'humanitarian crisis', after five battles between the sides resulted in the displacement of 159 people from the town of Rio Ciego. Some criminal gangs were also reported to have adopted guerrilla tactics, planting landmines to deter rivals and the security forces from entering their territory.

Criminal gangs also caused a diplomatic spat between Colombia and Venezuela. On 19 August, Venezuela's President Nicolás Maduro ordered the closure of a border crossing in the western state of Tachira after three of the country's soldiers were attacked by armed men. Maduro claimed that the troops had been ambushed by Colombian 'paramilitaries' who were transporting goods out of Venezuela. Although his government failed to provide more information on the attack, the groups he referred to were almost certainly Colombian criminal gangs, as both Los Urabeños and Los Rastrojos were known to operate in the area. (It was possible that the attacks were perpetrated by Venezuelan 'mega-bands', but these criminal groups were primarily concentrated in large urban centres.)

Despite the remaining obstacles in the peace process and the threat from a defiant ELN, Colombia drew ever closer to ending the Western Hemisphere's oldest insurgency. Bogota faced the challenge of implementing any agreements made in Havana while dealing with popular mistrust of FARC, as well as the financial constraints imposed by declining oil revenues. The Colombian economy weathered the fall of international commodity prices relatively well. Yet the government started in 2015 negotiations for new lines of international aid, especially those from Washington (which had provided US$10 billion in 15 years under Plan Colombia, a programme designed to aid counter-insurgency measures and tackle drug trafficking). Such international support was needed to implement Colombia's national and local efforts for developing areas affected by the conflict, and for reintegrating insurgent fighters into mainstream society.

Mexico

Key statistics	2014	2015
Conflict intensity:	High	High
Fatalities:	15,000	17,000
New IDPs:	9,000	
New refugees:	1,000	

Mexico's criminal underworld underwent significant changes in 2015, as a new, exceptionally violent drug cartel rose to challenge established powers, and Joaquin 'El Chapo' Guzmán, the country's most high-profile crime boss, escaped from a maximum-security prison. The heightened competition between gangs was reflected in the homicide rate, which grew continually between March and May – the first time since 2013 that it had risen for three consecutive months. Overall, there were more homicides in 2015 than in any year since 2011, reversing the gradual decline in violence that had allowed President Enrique Peña Nieto to focus on macroeconomic reforms. As a consequence, questions arose about the Mexican state's ability to deal with crime, particularly as the September 2014 disappearance of 43 trainee teachers in Iguala remained unresolved, and continued to provoke ire against the government. In western Mexico, the strength of the Sinaloa Cartel and the brutality of the ascendant Cártel de Jalisco Nueva Generación (CJNG) indicated the limited progress of government security initiatives, especially the reform of the national police structure and the proposed dissolution of municipal law-enforcement agencies.

Unstable criminal underworld

Between 2012 and 2014, Mexico experienced a gradual decline in the homicide rate. The reversal of this trend in 2015, with an 8.6% rise in homicides and around 17,000 homicides in total, was driven by violence in states affected by turf wars between armed groups. A case in point was the most violent state in 2015, Guerrero, which is located on the western coast and experienced 30% more homicides during the first ten months of the year than in the same period in 2014. Interim Governor of Guerrero Rogelio Ortega Martínez said in August that the shift resulted from ongoing battles for territory between the Beltrán Leyva Cartel, the Sinaloa Cartel and the CJNG, as well as a number of smaller local groups. Investigators from the general prosecutor's office also reported that members of two other major gangs, La Familia Michoacána and the Knights Templar, were operating in the state, albeit in relatively small numbers. However, no cartel was strong enough to dominate the drug trade.

Guerrero has been a focal point for criminal activity due to its lucrative opium market and the weakness of the state's presence there, which combine to generate intense competition between gangs. The disappearance of the trainee teachers in Iguala, a town in northern Guerrero, resulted in the criminal prosecution of local police officers and its former mayor. Ortega Martínez, who assumed power after his predecessor stepped down over the incident, stated in July that Guerrero's opium production rivalled that of Afghanistan (widely considered to be the world's top producer).

Although President Peña Nieto attempted to deal with the violence in the state, his approach differed little from past responses – both his own and those of his predecessor, Felipe Calderón. As in previous crises in Michoacan and Tamaulipas states, the government primarily relied on an increase in manpower, reinforcing Guerrero security personnel with 12,000 soldiers from December 2014 onwards. The following October, with the violence undiminished, Secretary of the Interior

Miguel Angel Osorio Chong announced that more federal forces, including a new anti-kidnapping unit, would be sent to protect the port of Acapulco, one of Mexico's main tourist destinations. The government also took deliberate steps to encourage economic growth in Guerrero, announcing that a highway would be built between the coast and the Tierra Caliente region.

Jalisco was another western state that received large numbers of federal reinforcements. Since 2010, its residents had witnessed the transformation of the CJNG, an extremely violent group, from a minor local actor into one of the most powerful gangs in Mexico. In order to curb the CJNG, the federal government launched *Operation Jalisco*, which backfired. On the day the operation began, 1 May 2015, the group staged a bold display of force against the state by downing a military helicopter, causing the deaths of four members of the Mexican Army and one police officer. Four more people died in the ensuing violence that engulfed almost all of Jalisco State, as well as neighbouring regions. The CJNG established 39 roadblocks – mainly in Jalisco but also in the neighbouring states of Guanajuato, Colima and Michoacan – while 11 banks and 16 petrol stations were set on fire. National Security Commissioner Monte Alejandro Rubido subsequently acknowledged that the CJNG had 'very significant firepower', adding that the group had destroyed the helicopter with an RPG-7, a Russian-made grenade launcher. Jalisco later received reinforcements of 6,000 more federal security personnel, including members of the Federal Police and the army.

The downing of the helicopter was the latest in the CJNG's series of direct attacks on government forces – a rare action by a Mexican criminal group, even the feared Los Zetas at the height of its power. Between mid-March and mid-April, the CJNG carried out several well-organised ambushes in western Jalisco that killed 28 people, most of them police officers. On 6 April, 15 officers died in a single ambush set up by suspected members of the CJNG, who burned vehicles to block a road used by the police.

Both Mexico and the United States expressed grave concern about the rise of the group. The Mexican general prosecutor's office reported in April 2015 that the CJNG had established a presence in eight states, as well as Mexico City, having only been in four in 2013. As a result, it had come to rival the Sinaloa Cartel, which the government estimated to have a presence in between six and nine states. While in its two decades of operations the Sinaloa Cartel had established a large, resilient network of international clients, a July 2015 report by the US Drug Enforcement Administration (DEA) referred to the latter group as the 'fastest growing [drug-trafficking organisation] in Mexico' and a major driver of the rise in violence in 2015. The DEA attributed the CJNG's growth to two main factors.

The first of these was the expansionist approach the group had adopted since its emergence, in 2011, as a splinter cell of the Milenio Cartel, then a local enforcer for the Sinaloa Cartel. The CJNG directly challenged Los Zetas, which at the time was the most brutal and militarily advanced gang in Mexico, and exploited 'the weaknesses of rival cartels to take over new territories', sparking turf wars. The second factor identified by the DEA was the deterioration of most other gangs, caused by fighting between them in the overcrowded underworld of western Mexico. Control of drug trafficking in Zacatecas State, for example, has been contested by the five main criminal groups present in Guerrero, as well as Los Zetas and the Gulf Cartel.

The general prosecutor's office identified nine main criminal groups, of which the most powerful were the CJNG and the Sinaloa Cartel. Although the latter had declined for several years, some Mexican media reports suggested that it slowly returned to growth in 2015, mainly due to expansion in Baja California Sur State. The seven other major gangs – the Juarez Cartel, the Tijuana Cartel, the Gulf Cartel, the Beltrán Leyva Cartel, Los Zetas, La Familia Michoacána and the Knights Templar – lost territory and armed capability. The simultaneous attrition of so many Mexican criminal groups was unprecedented.

Growing security challenges

The pressure on the Mexican government generated by the rising violence was compounded by challenges from civil society related to corruption in, and human-rights abuses by, the security forces – all of which cast doubt on the strength of state institutions. In Guerrero State, there were large-scale protests against alleged police corruption in relation to the 43 trainee teachers who had been missing since September 2014. Thousands marched in several major cities on 27 September 2015, drawing attention to the estimated 25,000 disappearances that had occurred in the preceding eight years.

The problems with Mexico's security apparatus became particularly apparent on 11 July 2015, when Guzmán, leader of the Sinaloa Cartel, escaped from the Altiplano maximum-security prison through a 1.5-kilometre tunnel leading from the shower room to a house on a construction site. The tunnel was equipped with ventilation, lights and a motorcycle on a rail. This was the second time Guzmán had escaped from a Mexican jail, having fled Puente Grande prison in 2001.

Following the escape, Chong revealed that security officers only reached Guzmán's cell 18 minutes after it was reported that he was no longer appearing on surveillance cameras. The following day, 22 people were placed under temporary arrest on suspicion of aiding Guzmán's escape. On 24 July, three employees of the prison were formally charged with 'facilitating the escape'. Four more were charged with the same offence on 7 September.

Guzmán was constantly on the run after this escape. Ismael 'El Mayo' Zambada, who had managed the Sinaloa Cartel during Guzmán's imprisonment, remained in charge of its day-to-day affairs. The group appeared to be stable under Zambada's leadership, as its strongholds in northwestern Mexico saw no significant increase in clashes with rival organisations.

One of the consequences of the chronic criminal violence in Mexico has been the rise of vigilante self-defence groups, which have become a third actor in the conflict between the gangs and state security forces. This development has been most apparent in western states, with such groups particularly active in Michoacan since 2013, despite government efforts to disarm them. The impact of vigilante organisations on state authority and human security in Michoacan became clear in November 2015, when the National Human Rights Commission

(CNDH) reported that they were present in 56% of the state's territory, or 33 municipalities. The CNDH added that the rights of citizens were being violated due to the absence of state authority in Michoacan. Making matters worse, the CJNG expanded into Michoacan in 2015. Following its established strategy, the cartel exploited the weakness of a rival – in this case the Knights Templar, previously the dominant force in the state.

In Michoacan, as in other Mexican states, there were reports that civilians were displaced by the increasingly complex competition between armed actors. The level of displacement in Mexico was comparable to that of countries engaged in civil wars: between 2009 and 2014, around 236,800 Mexicans left their homes to escape crime-related violence, according to a 2015 report by the government's official statistics agency. The publication of this figure was the government's clearest recognition of the magnitude of the problem. The high number reflected the strong emphasis on territorial control of the criminal groups, whose turf wars and use of some areas for illegal drug manufacture and trafficking were among the main drivers of forced displacement.

Reacting to the deteriorating public perception of security, President Peña Nieto announced new policy initiatives and a cabinet reshuffle, replacing the heads of seven ministries (while retaining Chong). The main change to the state-security leadership occurred at the National Security Commission, with Renato Sales Heredia succeeding Monte Alejandro Rubido Garcia as head of the organisation.

Peña Nieto's State of the Union address took place in September, marking the halfway point of his six-year term. Against a background of rising homicides, and with Mexicans concerned about high-profile cases of abuse and assassination, he acknowledged that the past year had been 'difficult'. The president identified the disappearances in Iguala, the cartel leader's flight and persistent allegations of corruption as having 'damaged the mood of Mexicans, as well as their confidence in government institutions', announcing that he would work with congress to draft new laws aiming at combating graft and other crime. He also proposed a bill for creating special economic zones in areas considered to be 'at risk' from violent crime, with the aim of creating jobs and improving productivity there.

Another change implemented after Guzmán's escape was the adoption of a more receptive approach to extradition requests from Washington. Accordingly, on 30 September, Mexico extradited 13 criminals to the US. They included US citizen Edgar Valdez Villarreal, known as 'La Barbie', and former Gulf Cartel leader Jorge Costilla Sánchez. This was a significant reorientation for the Peña Nieto administration, which had hitherto been more cautious and less cooperative than Calderón's government in dealings with Washington. Yet Peña Nieto's main security-reform initiative had not been fully implemented by the end of the year, despite having been announced in late 2014. Called *mando unico* (single command), the programme aimed to simplify the structure of the Mexican police, incorporating municipal forces under the umbrella of state governments and allowing for improved vetting and oversight of officers. The initiative faced resistance from sectors of congress and many local authorities. Although 71% of municipalities had agreed to implement the initiative, it faced political resistance in some of the most violent states in the country, including Guerrero.

Mexican non-governmental organisations, international human-rights groups and some parts of the media increasingly saw the government's failure to fully investigate the disappearances in Iguala as symptomatic of a state-wide crisis in which institutions were unable to provide security. The fate of the trainee teachers and the identities of those responsible remained unclear. The remains of only one of the missing were identified in December 2014 through DNA tests. The students were believed to have been taken into custody by local police on their way to a protest at an event hosted by Maria de los Angeles Pineda Villa, the wife of Iguala's mayor. They were allegedly handed to United Warriors criminal group, which had connections with some of Pineda Villa's relatives. According to the government's version of events, made public by the then-attorney general Jesús Murillo Karam, the trainee teachers were killed and their bodies burned in a rubbish dump in the neighbouring town of Cocula.

But an independent group of experts assembled by the Inter-American Commission on Human Rights (IACHR) found the government version inconsistent with developments on the ground, while observing that there were contradictions in accounts of purported witnesses to the killings, particularly regarding the manner in which the trainee teachers were executed. The experts agreed to remain in the country until April 2016, in order to continue their investigation.

In October 2015, Rose-Marie Belle Antoine, then head of the IACHR, said that 'the magnitude of the problem of disappeared persons in Mexico [was] alarming'. She added that the country suffered from a 'situation of extreme insecurity and violence, serious human rights violations, especially enforced disappearances, extrajudicial executions, and torture'. The preliminary conclusions and statements of the IACHR were backed by the UN High Commissioner for Human Rights, Zeid Ra'ad al-Hussein, who also called for the swift withdrawal of the military from the streets, due to reports that they had carried out human-rights abuses against civilians. The US State Department subsequently announced that it would withhold US$5 million of US$148m in aid to Mexico – allocated under the Merida Initiative, designed to improve the country's security – due to lack of evidence that the Mexican government was addressing such human-rights problems.

By the end of the year, the Peña Nieto administration was still pushing for *mando unico,* its police-reform initiative, as the key means for addressing insecurity and human-rights abuses. However, many Mexicans believed that the programme was being implemented far too slowly, given the effects of widespread cartel violence and police corruption. In light of these concerns and growing international pressure to strengthen the rule of law, Peña Nieto shifted his focus towards human rights and criminal violence, at the expense of the macroeconomic reforms and rebuilding of Mexico's image abroad that he had favoured earlier in his presidency.

Chapter Nine

Explanatory Notes

The Armed Conflict Survey provides analysis and data on active armed conflicts involving states and non-state armed groups across the world. The data in the current edition is accurate according to IISS assessments as at January 2016, unless specified. Inclusion of a territory, country, state or group in *The Armed Conflict Survey* does not imply legal recognition or indicate support for that entity.

General Arrangement and Contents

The introduction provides an overview of some of the key themes of the book. Next, four analytical essays focus on select cross-national trends: pessimism and planning in armed conflict; the political economy of violent conflict; the threat from foreign fighters; and the shifting roles of female combatants. A graphical section follows, presenting a number of conflict maps and graphics. The sections providing data and analysis on each conflict are organised according to region. The book closes with a reference section.

The Chart of Conflict

The Chart of Conflict shows the location and intensity of armed conflicts and terrorist attacks. The theme of the 2016 edition is armed conflict in urban centres.

Individual conflict entries in *The Armed Conflict Survey*

For each conflict, there is an analytical essay identifying and critically assessing key developments in the political, military and humanitarian aspects of the conflict. This is complemented by an analysis of conflict intensity and data on fatalities, refugees and IDPs.

Conflict intensity

For each active conflict, an assessment is made as to the intensity of the fighting:

- High-intensity conflicts involve frequent (daily) armed clashes between governments, government forces and insurgents, or among non-state armed groups that control territory. Typically, fatalities as a direct result of conflict exceed 3,000 per year.
- Medium-intensity conflicts involve regular armed clashes between governments, government forces and insurgents, or among other non-state armed groups that control territory. Typically, fatalities as a direct result of conflict exceed 300 per year.
- Low-intensity conflicts involve occasional clashes between governments, government forces and insurgents, or among other non-state armed groups that control territory. Fatalities as a direct result of conflict are typically below 300 per year.

Fatalities

Fatality statistics in *The Armed Conflict Survey* relate to military, insurgent and civilian lives lost as a direct result of armed conflict. Some data is estimated and all results are rounded. The figures are derived from publicly available information, including international, national and local press reports, as well as the annual assessments of international organisations (where available). Care is taken to ensure that our data is as accurate and free from bias as possible. The data reflect judgments based on information available to the IISS at the time the book is compiled. It may be subject to revision.

Refugees

According to the 1951 Refugee Convention (Article 1A(2)), the term 'refugee' is applied to a person who, 'owing to well-founded fear of being persecuted for reasons of race, religion, nationality, membership of a particular social group or political opinion, is outside the country of his[/her] nationality and is unable or, owing to such fear, is unwilling to avail himself[/herself] of the protection of that country; or who, not having a nationality and being outside the country of his[/her] former habitual residence as a result of such events, is unable or, owing to such fear, is unwilling to return to it'.

The refugees entry refers to the number of people that as a result of conflict moved out of the host country in the last year. It measures the flow rather than the total number, or stock, of refugees created by a conflict.

Although multiple sources are consulted as part of our analytical process, refugee data used in *The Armed Conflict Survey* is sourced from the United Nations High Commissioner for Refugees.

Internally displaced persons (IDPs)

IDPs are individuals or groups who have been forced or obliged to flee their homes as a result of, or in order to avoid, the effects of armed conflict, situations of generalised violence or violations of human rights. The IDP figures in the tables correspond to the number of people displaced that particular year, and not to the total number of IDPs since the beginning of the conflict. IDP figures and information come from the Norwegian Refugee Council's Internal Displacement Monitoring Centre.

Rounding

Figures have been rounded as follows: 1–12 none; 15–60 to the nearest five; 60–100 to the nearest ten; 100–200 to the nearest 25; 200–500 to the nearest 50; 500–1,000 to the nearest 100; 1,000–2,500 to the nearest 250; 2,500–10,000 to the nearest 500; 10,000–50,000 to the nearest 1,000; 50,000–200,000 to the nearest 5,000; and 200,000 plus to the nearest 10,000.

Attribution and acknowledgements

The IISS owes no allegiance to any government, group of governments, or any political or other organisation. Its assessments are its own, based on the material available to it from a wide variety of sources. Some data in *The Armed Conflict Survey* are estimated. Care is taken to ensure that these data are as accurate and free from bias as possible. The Director-General and Chief Executive and staff of the Institute assume full responsibility for the data and judgements in this book. Comments and suggestions on the data and textual material contained within the book, as well as on the style and presentation of data, are welcomed and should be communicated to the Director of Editorial at: IISS, 13–15 Arundel Street, London WC2R 3DX, UK; email: publications@iiss.org. Application to reproduce limited amounts of data may be made to the publisher: Taylor & Francis, 4 Park Square, Milton Park, Abingdon, Oxon, OX14 4RN; email: society.permissions@tandf.co.uk. Unauthorised use of data from *The Armed Conflict Survey* will be subject to legal action.

Key to regional maps

Status and types of conflict

High-intensity – frequent (daily) armed clashes between governments, government forces and insurgents, or among non-state armed groups that control territory.

Medium-intensity – regular armed clashes between governments, government forces and insurgents, or among other non-state armed groups that control territory.

Low-intensity – occasional clashes between governments, government forces and insurgents, or among other non-state armed groups that control territory.

Non-insurgent unrest

Ceasefire/Amnesty – agreed and/or announced by recognised leaders of parties to the conflict. Does not stand as a resolution of the conflict and does not suggest that all conflict has stopped.

Peace accord – formal resolutions of conflict ratified by recognised leaders of parties to the conflict or between intervening party and post-conflict government. In some cases, conflict may still persist.

Index

N

T

U